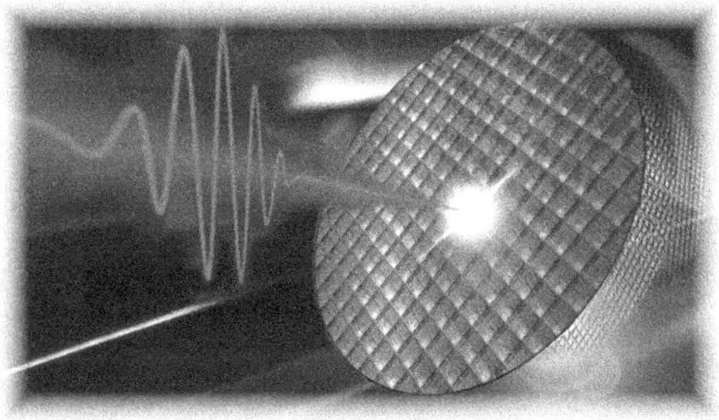

INDUSTRIAL APPLICATIONS OF ULTRAFAST LASERS

World Scientific Series in Materials and Energy

ISSN: 2335-6596

Energy and sustainability are keywords driving current science and technology. Concerns about the environment and the supply of fossil fuel have driven researchers to explore technological solutions seeking alternative means of energy supply and storage. New materials and material structures are at the very core of this research endeavor. The search for cleaner, cheaper, smaller and more efficient energy technologies is intimately connected to the discovery and the development of new materials.

This collection focuses onmaterials-based solutions to the energy problem through a series of case studies illustrating advances in energy-related materials research. The research studies employ creativity, discovery, rationale design and improvement of the physical and chemical properties of materials leading to new paradigms for competitive energy-production. The challenge tests both our fundamental understanding of material and our ability to manipulate and reconfigure materials into practical and useful configurations. Invariably these materials issues arise at the nano-scale!

For electricity generation, dramatic breakthroughs are taking place in the fields of solar cells and fuel cells, the former giving rise to entirely new classes of semiconductors; the latter testing our knowledge of the behavior of ionic transport through a solid medium. Inenergy-storage exciting developments are emerging from the fields of rechargeable batteries and hydrogen storage. On the horizon are breakthroughs in thermoelectrics, high temperature superconductivity, and power generation. Still to emerge are the harnessing of systems that mimic nature, ranging from fusion, as in the sun, to photosynthesis, nature's photovoltaic. All of these approaches represent a body of materials–based research employing the most sophisticated experimental and theoretical techniques dedicated to a commongoal. The aim of this series is to capture these advances, through a collection of volumes authored by leading physicists, chemists, biologists and engineers that represent the forefront of energy-related materials research.

Published

Vol. 11 *Industrial Applications of Ultrafast Lasers*
by Richard Haight (IBM TJ Watson Research Center, USA) and
Adra V. Carr (IBM TJ Watson Research Center, USA)

Vol. 10 *World Scientific Reference on Spin in Organics*
(In 4 Volumes)
edited by Zeev Valy Vardeny (University of Utah, USA) and
Markus Wohlgenannt (University of Iowa, USA)

Vol. 9 *Conjugated Polymers and Oligomers: Structural and Soft Matter Aspects*
edited by Matti Knaapila (Technical University of Denmark, Denmark)

For further details, please visit: http://www.worldscientific.com/series/mae

(Continued at the end of the book)

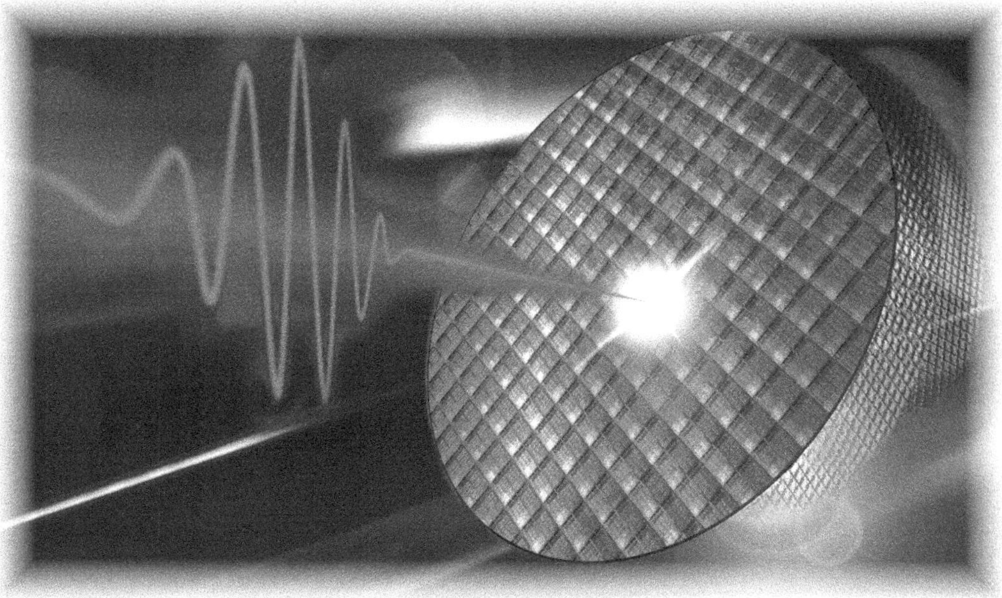

MATERIALS AND ENERGY – Vol.11

INDUSTRIAL APPLICATIONS OF ULTRAFAST LASERS

Richard Haight • Adra V. Carr

IBM T.J. Watson Research Center, USA

World Scientific

NEW JERSEY • LONDON • SINGAPORE • BEIJING • SHANGHAI • HONG KONG • TAIPEI • CHENNAI • TOKYO

Published by

World Scientific Publishing Co. Pte. Ltd.

5 Toh Tuck Link, Singapore 596224

USA office: 27 Warren Street, Suite 401-402, Hackensack, NJ 07601

UK office: 57 Shelton Street, Covent Garden, London WC2H 9HE

Library of Congress Cataloging-in-Publication Data
Names: Haight, Richard, author. | Carr, Adra V., author.
Title: Industrial applications of ultrafast lasers / Richard Haight (IBM TJ Watson Research Center, USA), Adra V. Carr (IBM TJ Watson Research Center, USA).
Other titles: World Scientific series in materials and energy ; v. 11.
Description: Singapore ; Hackensack, NJ : World Scientific, [2018] | Series: Materials and energy, ISSN 2335-6596 ; vol. 11 | Includes bibliographical references.
Identifiers: LCCN 2017054508| ISBN 9789814569002 (hardcover ; alk. paper) | ISBN 9814569003 (hardcover ; alk. paper)
Subjects: LCSH: Lasers--Industrial applications. | Laser pulses, Ultrashort.
Classification: LCC TA1677 .H36 2018 | DDC 621.36/6--dc23
LC record available at https://lccn.loc.gov/2017054508

British Library Cataloguing-in-Publication Data
A catalogue record for this book is available from the British Library.

For any available supplementary material, please visit
http://www.worldscientific.com/worldscibooks/10.1142/8985#t=suppl

Typeset by Stallion Press
Email: enquiries@stallionpress.com

RH: *This book is dedicated to my mother, Lucille, who insisted that education and the pursuit of knowledge was and should be a guiding light in the life of her children. I am also filled with gratitude for the support of my wife Judy and my children Elliott and Benjamin for making every day a rich, challenging, and ultimately fulfilling journey.*

AVC: *For Yancey, who has always been my source of sanity, support and much-needed coffee breaks.*

Contents

Foreword

Scientific study spanning decades of experimental work contributed to the development and contents of this book. Much of the work described here was carried out at the authors' host institutions; IBM's TJ Watson Research Center in Yorktown Heights, N.Y. as well as at the University of Colorado in Boulder. While there have been a significant number of excellent books written about the development of lasers in general and more specifically femtosecond lasers, in recent years the applications of these unique optical tools to solve industrial problems have gained momentum. One might suppose that the somewhat academic, almost esoteric research into physical phenomena at femtosecond and attosecond timescales, realized through the application of ultrashort pulsed lasers, would not necessarily find utility in solving technical problems confronting our most advanced industries. What is becoming clear is that an increasing number of industrial companies, whose focus is in areas that include semiconductor chip manufacturing, automobiles, communications, materials and more, and the companies that support these industries, are requiring ever more sophisticated approaches to studying and analyzing their materials, products, and processing methods. Chip manufacturing, for example, has employed increasingly exotic materials at ever smaller spatial dimensions, now well into the nanoscale, that require new approaches to analyze their behavior and properties at both ultra-small and ultrashort timescales. What are the electrons doing in these materials? What types of thermal transport affect the material properties during processing and manufacture? How can these materials be machined with the least amount of damage adjacent to the altered area? How can materials be transformed, precisely controlled, and imaged on ultrashort time scales so as to observe transient behavior? There are, in many cases, more questions than answers.

This book attempts to capture and address some of these compelling questions in detail, and in other cases to survey the wide range of ongoing work on issues relevant to industry.

There are, of course, many people to thank that have contributed to work discussed in this book. In the area of femtosecond photomask repair, considerable thanks and gratitude are extended to Al Wagner and Pete Longo. Together we built the first femtosecond photomask repair tools installed in the IBM-Burlington Mask House. The tool specifically discussed in this book, called Mask Advanced Repair System (MARS) was installed and ran successfully for many years in the Mask House; the technology was eventually licensed to a company that sells these tools worldwide. Great credit and appreciation goes out to the numerous students, postdocs, visitors, and collaborators whose diligent efforts contributed so importantly to the many experiments and insights, and published papers over the years. Additional thanks to Margaret Murnane and Henry Kapteyn at the University of Colorado at Boulder and JILA for close to 30 years of collaboration and fruitful discussions on femtosecond laser technology and the science of high-harmonic generation. A heartfelt thanks to Len Feldman at Rutgers University for years of mentorship, collaboration, and friendship dating back to 1981 and to Zvi Ruder at WSPC, executive publisher for many fruitful discussions and his continued support of this work.

Richard Haight
Adra V. Carr
Yorktown Heights, NY (2017)

Chapter 1

Introduction

What makes an ultrashort laser pulse so useful?

As almost everyone would agree, lasers are extraordinarily useful devices whose output has found applications in laboratory, commercial, and military realms. Continuous wave or CW lasers are ubiquitous and are used in music and video players, laser light shows, and fiber-based communications. Diode lasers can be modulated simply by turning the device on and off, or via an outside modulator to encode information. Highly stabilized CW lasers have provided the best clocks currently devised by man in the form of *atomic clocks*. They occupy the heart of our satellite-based global positioning systems (GPS), used by nearly everyone who drives an automobile, boat, or a plane. CW lasers and their applications have been addressed in great depth in literature and will only be discussed in this book as they relate to pulsed laser systems and their applications.

But we still have not answered the question posed above: What makes an ultrashort pulse so useful? Let us briefly discuss pulsed lasers in general. Pulsed lasers are just that; they squeeze the light they generate into bunches or pulses that contain, in many cases, orders of magnitude more photons per unit time than those produced by simply dicing up the output of a CW laser with, for example, a simple mechanical shutter or "chopper". The output of modern pulsed lasers can be quite intense and are often used to melt, machine, or vaporize materials in both laboratory and industrial applications. One example of a widely used pulsed laser is the workhorse Q-switched neodymium-doped yttrium aluminum garnett (Nd:YAlG) laser that can produce extremely intense nanosecond and picosecond pulses in the infrared at 1,064 nm. Frequency up-conversion to the green at 532 nm, ultraviolet (UV) at 355 nm, or even the deep UV at 266 nm can be achieved in nonlinear crystals with great efficiency. A similarly useful gas-based excimer laser can produce joules of nanosecond UV light pulses that are routinely used in lithographic printing of today's integrated circuitry. Linewidth-narrowed 193 nm ArF excimer lasers are used to produce leading-edge nanometer-scale features on Si wafers coated with chemically amplified photoresists and are, at present, the only method for mass production of monolithic integrated circuits in the world. The rate at which chips are manufactured in a modern semiconductor fab, utilizing pulsed excimer lasers for lithography is unrivaled by any other printing method such as electron-beam or extreme UV (EUV).

But, in some sense, we still have not addressed the difference between the broad category of pulsed lasers and *ultrashort*, or *femtosecond*, pulsed lasers. We are assisted in this endeavor by evaluating the dynamic properties of materials (both gases and solids) that act as a guide, where response to an impulsive excitation varies dramatically depending on the timescale of the excitation.

Consider, for example, the interaction of an intense pulse of laser light with a solid material. Visible photons in the light pulse are absorbed by electrons in the material and raised to normally unoccupied excited states of the system. If the pulse intensity is weak, those electrons will relax back to their ground state through the emission of phonons; the main consequence of this is heating of the material. The fundamental timescale for electron–phonon scattering in a metal is a few picoseconds, while for a semiconductor, it is a few hundred femtoseconds. It is similarly short in insulators. For all of these materials under pulsed visible or UV light illumination, it may take many electron–phonon interactions to relax the electronic system to its ground state with the attendant production of phonons that heat the solid.

Now consider a far more intense pulse of laser light such that the high number of photoexcited electrons produces an even higher number of phonons. For a sufficiently intense pulse of light, the temperature may rise many hundreds of degrees or more, enough to melt the material. If the pulsewidth of the light is relatively short, but not ultrashort, say hundreds of picoseconds to nanoseconds or more, the material will simply melt. Laser melting has been studied a great deal in the literature.

Melting may be the desired goal in a laser–solid interaction. An example of this, used widely in semiconductor manufacturing, is laser annealing of a crystalline semiconductor implanted with dopant atoms. The implantation of dopant atoms leaves the crystal in a disordered, amorphous state due to the collisions of the ions in the beam with the substrate atoms. To recrystallize the semiconductor, lattice annealing in a furnace allows the atoms to reorganize back to their crystalline state. During this time, dopant atoms can diffuse and occupy substitutional sites in the lattice; these dopant atoms are now "activated" in that they either donate to or accept an electron from the host lattice creating a population of electrons or holes. Laser annealing with relatively short pulses serves to activate the dopant atoms while limiting their diffusion within the lattice, thereby spatially constraining the activated region. As mentioned, activation refers to the replacement of the host atom with the dopant resulting in donation (n-type), or acceptance (p-type) of an electron within the host semiconductor.

If we consider an even more intense pulse that is short, but not ultrashort, laser-induced evaporation of the material occurs. This evaporation can be used to achieve a number of goals; the evaporated material may be deposited onto another substrate nearby for the purpose of building up layers of one material atop another. This is known as pulsed laser deposition (PLD). Or the evaporation of the material may be used to drill holes in the material for the purpose of machining it at the nanoscale, beyond the limits of standard machining tools and bits. Even higher energy pulses can result in the formation of a plasma whose light emission reveals the nature and composition of the material being ablated. This process is known as laser-induced breakdown spectroscopy (LIBS). Yet another application is to generate a pulse of broadband short wavelength radiation that can be collected and refocused for carrying out lithographic patterning of semiconductor chips (also called EUV lithography).

Figure 1 provides a schematic representation of the time domains and the types of physical processes that typically occur when excited by lasers of a given pulsewidth. This schematic is germane to our introductory discussions.

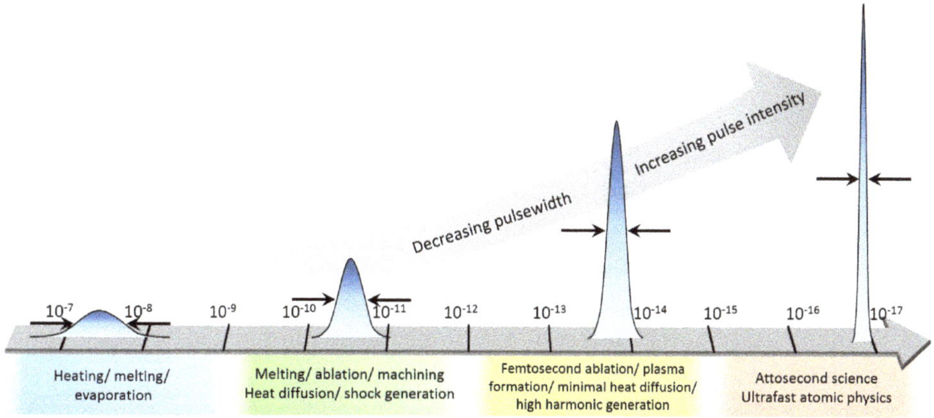

Fig. 1. Schematic description of the pulsewidth epochs, ranging from sub-microsecond to attosecond. With decreasing width, dramatic increases in the number of photons packed into a pulse are achieved.

All these processes can be achieved with a short ($\geq 10^{-12}$ s) but not ultrashort ($<10^{-12}$ s) pulse. In essence, such a pulse drives the conversion of the material sequentially from solid, to hot, to melted, and finally vaporized phases. Such a sequence is effective but can also present problems. For example, multi-picosecond to nanosecond pulsed laser ablation can produce splattered material and, since the process is violent on the spatial scale of the focused laser, shock waves can be produced. Shock waves and heat diffusion can result in significant extended damage of the material well beyond the focal region. While this is not necessarily a problem in spectroscopy of a material or low spatial resolution machining applications, it can be significant when using pulsed lasers to machine material in a finely controlled manner.

Ultrashort or ultrafast laser pulses possess a time structure whereby an extraordinarily large number of photons (i.e. energy) can be deposited in a material in a timescale below that associated with movement of atoms and ions. This means that within the focus of an intense femtosecond pulse of light, bonds can be broken at a rate that precludes thermal heating and melting of the material; for sufficiently short pulses, the system proceeds directly to a dissociated state allowing for surgical removal of material.

Another compelling effect associated with the focus of intense femtosecond pulses is the attainment of extraordinary intensities, well in excess of 10^{15} watts/cm^2. At these intensities, atoms are ionized to high levels and the electric field strengths within the focus are on the order of volts/Å resulting in the capture and manipulation of electrons. Among the many physical effects operative, the process of high-harmonic generation can occur, wherein multiples (typically odd but under the appropriate conditions even multiples are possible) of the driving laser frequency can be generated and harvested. Through the process of high-harmonic generation, conversion of light from infrared wavelengths to UV, vacuum UV, soft X-ray, and beyond can be achieved, all with pulsewidths that range from femtoseconds to attoseconds (10^{-18} s). Attosecond (and even zeptosecond) pulses, recently generated in laboratories possessing high powered femtosecond laser systems, are the shortest time phenomena ever created by man. Such short pulses have facilitated new studies in ultrafast atomic physics and have led to studies of the fundamental light–electron interactions in the photoemission of electrons from atoms and solids.

In diverse areas that include high-resolution machining of materials, semiconductor and atomic science, and medical science, employing femtosecond pulsed lasers has led to the forging of new areas of study and applications. In particular, industrial utilization of femtosecond light pulses has found applications in the areas of semiconductor science and technology through characterization, micro- and nanomachining, imaging, and spectroscopies. In this book, we describe some of the key applications of femtosecond lasers to these industrial uses.

The layout of this book beyond this brief Chapter 1, Introduction begins with Chapter 2 entitled *Ultrafast Laser Systems*. In this chapter, we present an in-depth discussion of the present state-of-the-art femtosecond amplified laser systems. Nearly all ultrafast amplified systems begin with a Kerr-Lens Modelocked (KLM), titanium-doped sapphire crystal oscillator that seeds an amplifier. The shortened term for these lasers is Ti:sapphire, which routinely produce the shortest near-infrared (~800 nm) pulses, down to a few femtoseconds. A somewhat different approach to generating femtosecond pulses involves the use of optical fiber-based lasers, although these systems tend to produce pulses in the 100 fs to sub-picosecond time range, with only recent progress yielding shorter pulselengths.

The fundamental basics of this operation will be discussed with an emphasis on their uses in an industrial environment. Present day systems are available from a number of commercial vendors in the US and Europe, and kits are available to construct modifications for particular applications. A common amplified femtosecond laser system begins with a Ti:sapphire based oscillator, typically driven by a diode pumped, intracavity-frequency-doubled CW neodymium-doped yttrium lithium fluoride (Nd:YLF) laser operating close to 530 nm, near the peak absorption of Ti:sapphire. The oscillator produces pulses at a 75–90 MHz repetition rate determined by the round-trip time of a pulse circulating within the oscillator cavity. For amplification, a single pulse from the oscillator is selected with an optical shutter called a "Pockels Cell". The selected, relatively weak pulse is then "regenerated" or amplified in a second oscillator powered by a lower repetition rate (1–10 kHz), but higher energy per pulse "pump" laser operating near 530 nm. A somewhat different implementation involves multipass amplification, wherein the oscillator seed pulse is passed multiple times through a pumped Ti:sapphire crystal; both schemes are designed to produce gains of $>10^6$ in energy/pulse. In addition to the generation of such short, intense pulses, delivery of these pulses will be described. Because of the short-pulsed nature of the light, care must be exercised in the types of optical components used. This includes the use of dielectric mirrors optimized for short pulses that can accommodate the larger bandwidth of the light. Nonlinear optical crystals must be made thin so as not to broaden the width of the input and output harmonic pulses. These are just a few examples of the issues confronted by the experimentalist when utilizing ultrashort pulses, and will be discussed.

Chapter 3 entitled *High-Harmonic Generation with Femtosecond Light* covers the physics and applications of high-harmonic generation. Nonlinear conversion of infrared and visible photons to shorter wavelengths awaited the invention of lasers where intense, coherent light was required to drive higher-order processes in crystalline materials. Initial studies date back to the early 1960s. Frequency doubling in non-centrosymmetric crystals utilizing the intense pulsed light from modelocked lasers became substantially more efficient and routine as laser pulsewidths shortened.

Reaching into the UV, vacuum UV, and soft X-ray regime required harmonic conversion in gases. Vacuum UV light at 118 nm was first generated in Xe gas in the early 1970s using picosecond lasers. In the 1980s, intense femtosecond pulses were employed to generate high harmonics in rare gases. A comb of odd multiple harmonics stretched well into the soft X-ray regime down to 7.5 nm. Further improvements in experimental apparatus and increases in the understanding of the high-harmonic generation process led to phase matching between the driving laser and the generated harmonics. This produced more intense pulses of light at ever shorter wavelengths down to, and below, the water window between the carbon absorption edge at 282 eV and oxygen at 530 eV.

Not only did high-harmonic generation produce spectrally pure pulses of light that were quasi-tunable from the vacuum UV to the soft X-ray region of the spectrum, but it made possible the generation of the first sub-femtosecond pulses. In this chapter, we will discuss the methods for generating high harmonics of a visible laser and methods to deliver and utilize this extraordinarily useful light.

In Chapter 4 entitled *Femtosecond Photoelectron Spectroscopy: Fundamentals and Electron Dynamics*, we describe one of the most successful applications of high-harmonic light in the realm of spectroscopy, and, in particular, photoelectron spectroscopy. Photoelectron spectroscopy is such an important and successful tool for studying both the electronic structure and chemical properties of materials; as an example, multi-user synchrotron light source facilities, costing hundreds of millions to billions of dollars to build and maintain, dot the landscape in the US, Europe, and Asia. To date, there are more than 90 such facilities worldwide.

While the wavelength range available at these facilities exceeds that available from femtosecond laser-generated high harmonics, the femto- and attosecond structure of the light achievable with lasers opens entirely new vistas of time-resolved physics of materials and gases. The light from high harmonics is spectrally pure, highly polarized, and tunable, leading to an unprecedented array of experiments on industrially important materials.

Chapter 5 entitled *Femtosecond Ablation and Mask Repair* describes in detail one of the fundamental uses of intense femtosecond laser pulses: micro- and nanoscale machining of materials through the process of ablation. Unique to femtosecond pulses is the ability to directly transform materials from the solid state to that of an energetic plasma, essentially bypassing melting. In doing so, the splatter of molten material and attendant shock-wave formation and heat diffusion are minimized, leading to surgical removal of material.

A significant portion of this chapter is devoted to the industrial application of femtosecond photomask repair. This quintessential application of femtosecond micro/nanomachining provides a unique instructional example of the benefits of ultrafast laser ablation. Removal of unwanted metal absorber material (defects) at the sub-micron level from a fused silica photomask substrate is an exacting process. While nanosecond pulses used for defect removal can splatter molten material and damage the underlying fused silica, femtosecond pulses eliminate these deleterious effects. In addition to defect removal, the addition of material through femtosecond-stimulated nonlinear deposition will also be described. Additional applications to laser eye surgery and transparent material machining will be presented as well.

In Chapter 6 entitled: *Femtosecond Photovoltage Spectroscopy and Device Physics*, we will describe fundamental and industrial applications of high harmonics and photoelectron spectroscopy. In the particular areas of industrial applications, we describe the extremely useful technique of femtosecond photovoltage spectroscopy (FPS). Inducing photovoltage shifts in semiconductor materials has been known and utilized in experiments dating back to the 1970s, where intensity-modulated light from a discharge lamp could be used to generate populations of electrons and holes in semiconductors through optical absorption. The presence of the electrons and holes would induce small changes to the internal dipole fields within the band-bent depletion region of the semiconductor near surfaces and interfaces. Since discharge lamps are a weak source of light, this modulation in band bending was small, and substantial modeling was required to evaluate the nature of the internal fields. With the advent of high-intensity, femtosecond light pulses, sufficiently dense electron-hole populations could now be generated to completely nullify the existing internal depletion field and fully flatten the bands of semiconductors whose doping levels ranged from intrinsic to $10^{18}/\text{cm}^3$ and beyond. Combined with photoelectron spectroscopy, buried layers of semiconductors under oxides and metals could now be studied, and a complete picture of the electronic structure of these complex stacks developed. Discussion of industrially relevant semiconductors, metal-oxide-semiconductor (MOS) structures, photovoltaic materials, and the underlying semiconductor physics will be presented.

In the final Chapter 7, entitled: *Survey of Ultrafast Light Sources, Applications, and Final Thoughts*, applications of femtosecond light pulses to EUV imaging and metrology, as well as a survey of additional applications will be presented. With the ability to generate substantial fluxes of light well in to the soft X-ray, applications of imaging and metrology to industrially relevant structures are emerging, as technologies such as EUV lithography are just now entering into production, driving the smallest features in integrated circuit printing to below 7 nm. We will survey many of these applications to provide a more complete view of the wide array of industrially relevant uses of ultrafast laser light. In addition, many new types of light sources have come online worldwide. These include X-ray free-electron lasers that operate in the Å wavelength regime, producing pulses into the femtosecond range, and next-generation ultrahigh brightness synchrotron facilities.

We hope that the descriptions within these pages provide a fundamental picture of the extraordinary uses and applications of ultrashort laser pulses and the host of compelling problems and challenges they can solve.

Chapter 2

Ultrafast Laser Systems

2.1 Introduction

The ability to generate an ultrashort pulse of light is at the heart of this book. In this chapter, we will discuss the basic laser systems, from oscillator to amplifier, that will provide the intense femtosecond and even attosecond pulses that will be utilized for applications ranging from material ablation to high-harmonic generation, electron spectroscopies, and more. While we will leave detailed discussion of the applications of femtosecond lasers to subsequent chapters, it is important to understand ultrashort pulse generation mechanisms and common ultrafast laser configurations. The potential for industrial applications has spurred the need for robust, powerful, and simple laser systems that take the laser outside of the university environment. Within the following sections, a basic understanding of the principles of a laser are assumed (for more references, see ([1, 2]), as we will only focus on the implementations and special concepts needed to understand *ultrafast* lasers.

In general, the laser system typically utilized in an industrial environment can be classified into two components: an *oscillator* and an *amplifier*. The *oscillator* can be thought of as our base laser cavity, responsible for outputting a train of pulses in the femtosecond regime. An overview of the physical mechanisms responsible for this operation including modelocking [1] and, more commonly, Kerr-Lens Modelocking (KLM) [3–5] can be found starting in Section 2.2.3. The *amplifier* is responsible for taking the output of the laser oscillator and increasing the optical power, ideally, without sacrificing the short pulselength of the incoming pulse. Multiple approaches have been used in amplification that have proved robust enough for industrial applications and can be found in Section 2.5.

In its most basic form (Fig. 1), a typical laser oscillator consists of a gain medium and cavity, pumped by an external laser [1]. The cavity is

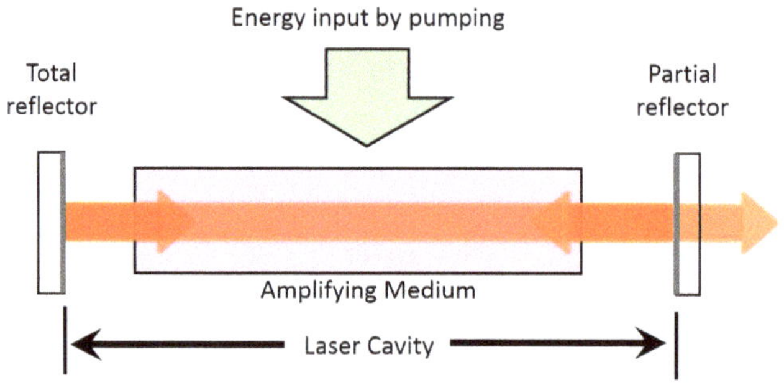

Fig. 1. Basic laser cavity showing a high-reflecting mirror, gain medium, and partial reflector that allows a fraction of the circulating light to emerge from the laser cavity. The gain medium as discussed in the text is Ti:sapphire, with the medium being pumped by an external light source.

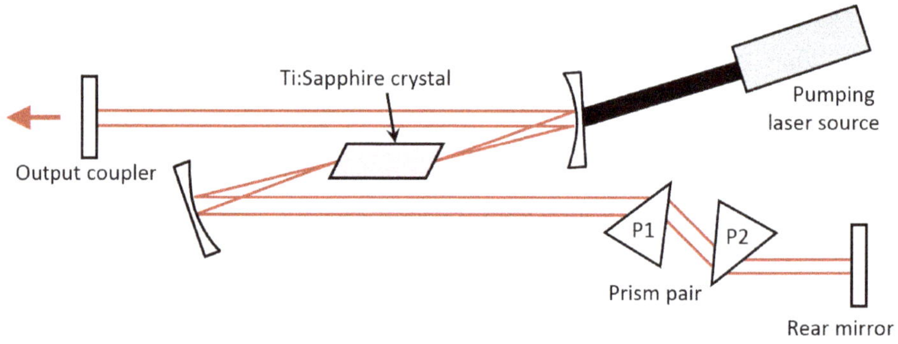

Fig. 2. Typical dispersion-compensated Ti:sapphire oscillator. A pumping laser source, typically an intracavity frequency-doubled Nd:YLF laser photoexcites the Ti:sapphire gain medium residing between curved mirrors. A prism pair (usually fused silica) is utilized to control dispersion within the cavity and optimizes pulsewidth, as discussed in the text.

bounded by two mirrors, one highly reflective mirror and one that is partially transmitting so that some of the light can escape for use outside the oscillator. The round-trip time is given by the distance traveled by a photon divided by the velocity of light in air. The gain medium in almost all of our discussions is a crystal of titanium-doped sapphire (Al_2O_3), more commonly known as Ti:sapphire [6, 7] (Fig. 2), and will be discussed in more detail in Section 2.3. As photons emitted from the gain medium circulate within the cavity and pass back into the crystal, they stimulate additional emission from the optically excited gain medium. This results in exponential growth of the photon population within the cavity, better known as lasing. The laser light circulating within this cavity is continuous in nature and not pulsed.

Pulsed operation requires a few additional constraints on the laser cavity, as we now discuss.

Early demonstrations of ultrashort pulse lasers relied on the concept of Q-switching [8, 9], whereby a modulating element was placed inside the laser cavity to precisely control within-cavity losses (or "Q" of the cavity). With the switching element in the cavity set for high loss, the population inversion in the lasing medium is allowed to grow, free from stimulated emission from the cavity. With the switching element set to low loss, this built-up energy is released in a single pulse. This design results in pulse lengths down to only several nanosecond, with it being limited by the photon round-trip travel time within the cavity. Soon after the initial demonstration of nanosecond pulses [10], picosecond pulses were made possible via employing the concept of modelocking [11]. It is this approach, and its special implementations, that have reduced these pulselengths to the femtosecond regime and is the basis for all major ultrafast lasers used in the industrial setting.

2.2 Modelocking

Modelocking can best be understood by considering the output of the laser cavity in the temporal domain. In a simple laser, many different wavelengths satisfy the resonance condition and can propagate within the cavity. If one was to "lock" the phases of the different cavity modes so they have a fixed phase relationship (see Fig. 3), this results in a periodic constructive interference of the output intensity as a function of time; this leads to the formation of a *pulse*. The period of this constructive interference, or repetition rate, corresponds to the round-trip travel time within the cavity. Fourier analysis also tells us that the temporal width of the features (the pulselength) is inversely proportional to the frequency bandwidth of the resonating wavelengths. Therefore, only a laser medium that is able to lase over a wide range of wavelengths (i.e. a large gain bandwidth) has the ability to produce very short pulses. Commonly utilized Ti:sapphire possesses a sufficiently large bandwidth [12] ranging from $\sim$650 nm to 1100 nm, to support pulses down to a few femtoseconds

In reality, there is no "natural" fixed phase relationship between different cavity modes. This phase relationship, therefore, can be imposed in two ways: active modelocking and passive modelocking.

2.2.1 *Active Modelocking*

Active modelocking is accomplished by periodically modulating cavity losses or the round-trip phase change in the cavity. Usually done through a physical

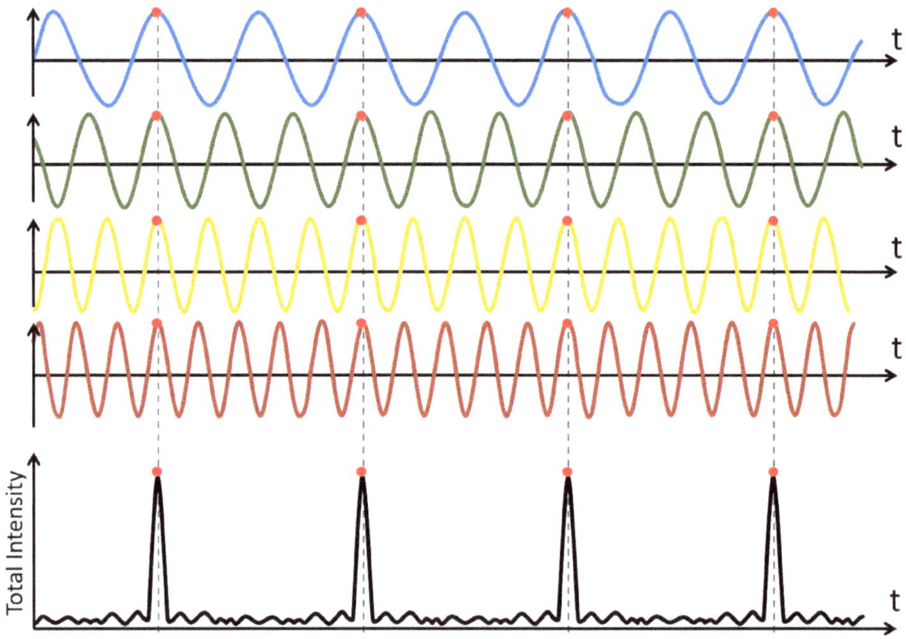

Fig. 3. Illustration of modelocking, where the "locked" phase relationship between wavelengths results in periodic constructive interference of the output intensity with time.

element such as an acousto-optic or electro-optic modulator, induced losses that are synchronized with the resonator round-trips produce output pulses when the cavity losses are low. In the case of acousto-optic and electro-optic modulators, these behave as shutters that change the lasing properties of the cavity. Since the round-trip time of a typical oscillator may be as short as 10–15 ns, there is no mechanical method available that can modulate optical properties on this timescale. An acousto-optic modulator, also known as a Bragg cell, operates by diffracting the laser light from a sound wave propagating in a crystal through which the laser light passes. The sound waves are created with a piezoelectric transducer driven by an electrical signal. A schematic picture of such a modulator is shown in Fig. 4. Use of this device limits efficient lasing only for a pulse satisfying the appropriate timing and path conditions.

A second optical modulator is electro-optic in origin. In this device, a time-dependent electric field is imposed on a crystalline material such as, lithium niobate; also known as the Pockels effect, the applied electric field introduces a birefringence in the material. The effect is that the polarization of the light circulating within the cavity can be rotated by as much as 90° and when used in conjunction with polarization control optics such as a ½-wave

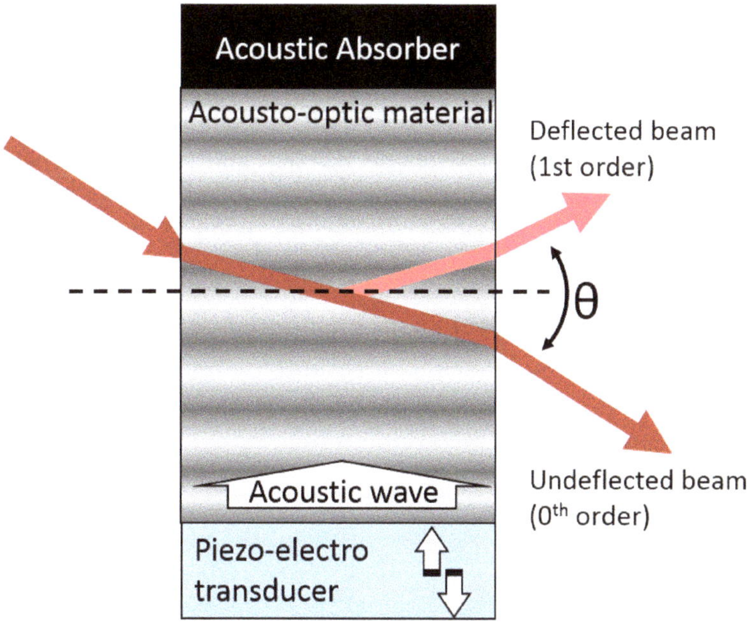

Fig. 4. Schematic of an acousto-optic modulator showing the diffraction at different angles θ of a laser pulse propagating through the device.

plate (which rotates the light polarization by 90°). An appropriately coated, fused silica optic oriented at Brewster's angle (typically ~56° for Ti:sapphire wavelengths) can be used to capture or reject light pulses from a cavity. Such an electro-optic element is called a Pockels Cell and is a critical component of regenerative amplifier cavities that will be described later in Section 2.4. The control of the cavity Q through the use of these modulators, or as we will see in Section 2.2.3 through a passive approach called "Kerr-lens modelocking", allows for temporal control of the power in the oscillator that results in the formation and circulation of stable pulses within the cavity.

The length of a pulse circulating within the laser cavity is heavily dictated by the speed at which the modelocking element can modulate, since the "wings" of the output pulse are determined by increase in extinction. This temporal narrowing leads to spectral broadening, as dictated by the Fourier relationship between time and wavelength (see Section 2.6). The gain profile for the gain medium in the cavity is usually a peaked function (producing more gain around a central wavelength), which then gives the effect of amplifying the central part of the resultant spectrum more strongly and giving narrower spectra. With these two competing forces in play, a steady state is reached after many round-trips of the pulse in the cavity, producing pulses with widths typically down to the picosecond regime.

The competition between intracavity spectral broadening mechanisms (like the broadening due to the gain medium) and the shortening by the modulator leads to a limit to how far this technique can be pushed for ultrashort pulse generation. Specifically, since the loss modulation function stays the same as the pulse shortens, it has less and less effect on shortening the pulse further as it circulates in the cavity.

2.2.2 *Passive Modelocking*

Passive modelocking is achieved by introducing a loss element into the cavity that is *intensity-dependent*, such as a saturable absorber material (Fig. 5). Saturable absorbers work by absorbing light at lower intensities, such as in

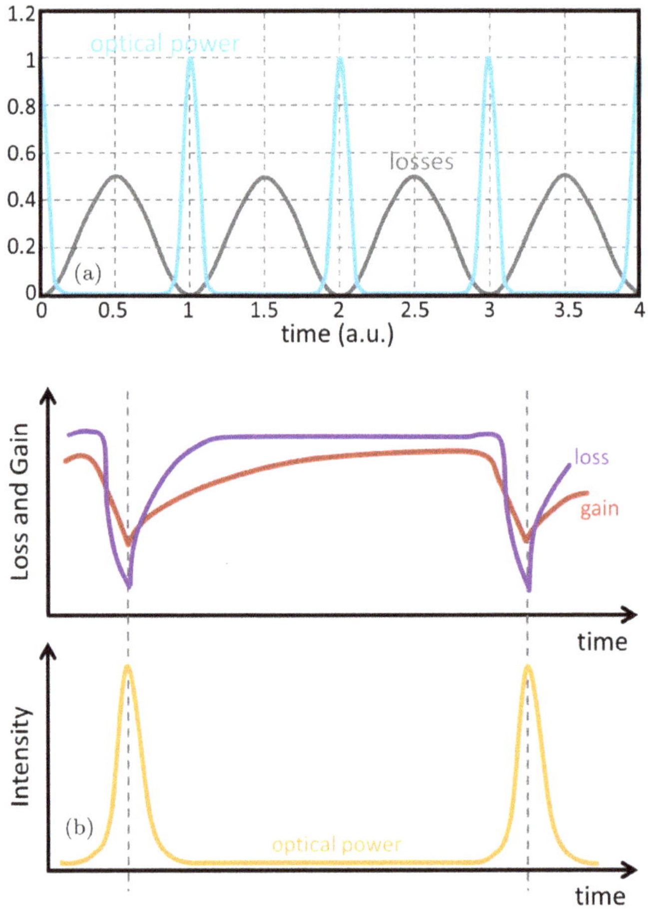

Fig. 5. Optical output power as a function of time for (a) active and (b) passive mode-locking using a saturable absorber. Corresponding cavity losses are also shown.

the leading or trailing edge of a pulse; at higher intensities near the center of the pulse, the material can no longer absorb light (it bleaches) and the pulse "punches" through. In this way, we are able to push past the limits of active modelocking since, as the pulse becomes shorter, the loss modulation becomes faster and is only limited by the recovery time of the material.

Ideally, as described above, the saturable absorber selectively absorbs low-intensity light (high loss) and transmits light of sufficiently high-intensity (low loss). Upon starting to circulate multiple modes within the laser cavity, the random phase relationship between modes will result in continuous-wave (CW) output with noisy fluctuations in the output intensity. If one of these intensity spikes is sufficient to saturate the absorber, the absorption will "bleach", decreasing the loss and allowing more gain within the cavity. Importantly, the peak of the spike will see low loss, while the lower power "wings" will be more attenuated, further shortening the pulse (as seen in Fig. 5). As with active modelocking, this shortening will occur with every pass of the pulse within the cavity until a steady state is reached. The primary limitations are due to both the response of the saturable absorber itself and the gain-bandwidth limit of the laser gain medium, since there is typically a peaked distribution of wavelengths that are able to be supported within the gain medium. Additionally, the large bandwidth of the ever-shortening pulse is increasingly affected by chromatic dispersion and nonlinearities in the gain medium as well. Even with these deleterious effects, this approach led to the first demonstrations of femtosecond-regime pulselengths [13].

2.2.3 *Kerr-Lens Modelocking*

Up until the 1990s, the ultrafast laser community was still relatively small, specialized, and exclusive to academia and national labs. The discovery of KLM in Ti:sapphire [14] was arguably the "kick-off" to the ultrafast revolution that ultimately resulted in sources stable and robust enough for industrial applications. Soon after initial demonstrations, pulselengths of less than 6 fs were reported directly out of the laser oscillator, without any need for additional pulse compression techniques [15].

KLM is a passive modelocking technique where the gain medium itself is responsible for the nonlinear intensity response, replacing the saturable absorber described in previous discussions. Let us consider the impact of the nonlinear index of refraction of the gain medium (n) as

$$n(I) = n_0 + n_2 I + \cdots . \quad (1)$$

Here, n_0 is the usual index of refraction, n_2 is the second-order index of refraction, and I is the instantaneous beam intensity in the gain crystal. Due to the Kerr effect on the incoming Gaussian laser beam (Gaussian optics is discussed in detail in Chapter 5), this nonlinear index of refraction, induced by the strong AC field of the light pulse, leads to a phase delay that is largest on the central beam axis and smallest on the outermost axis. The gain medium, therefore, behaves like a lens and leads to self-focusing of the beam as it propagates. To see how this self-focusing behaves and relates to other properties of our incoming beam, we can start by determining the phase delay between the on-axis and outermost positions in our beam. Considering the intensity of the Gaussian beam near the optical axis, we can make the approximation

$$I(r) = I_0 e^{-\frac{2r^2}{w^2}} \approx I_0 \left(1 - \frac{2r^2}{w^2}\right), \tag{2}$$

where I_0 is the beam's peak intensity, r is the distance from the optical axis, and w is the waist size of the beam. Relative to the on-axis phase ($\phi_{r=0}$, where ϕ is the phase), the relative phase at a distance r from the center of the Gaussian beam gives, after propagation through the laser medium of thickness d

$$\Delta\phi(r) = -2k_0 n_2 d\, I_0 \frac{r^2}{w^2}, \tag{3}$$

where k_0 is the free-space wavenumber of the laser. Recasting this approximate expression in terms of a thin focusing lens, this has the effective focal length of

$$f = \frac{w^2}{4n_2 d\, I_0}. \tag{4}$$

This result demonstrates that the degree of self-focusing within the laser gain medium is proportional to the instantaneous intensity within the crystal. Therefore, the higher the instantaneous intensity, the shorter the focal length and the more "extreme" the self-focusing effect. This effect alone is not enough to lead to the modulation of *loss* within the cavity that is needed for passive modelocking. An additional design feature needs to be employed to take advantage of this unique effect. One approach, called *hard-aperture KLM* [16], places a physical aperture in the beam that leads to lower losses when the Kerr-lens effect is stronger. The second approach, and more common, is the so-called *soft-aperture KLM* [4], where alignment of the cavity gives a larger beam size in the gain medium for lower intensity non-pulsed operation. Utilized with a tightly focused pump laser onto the crystal, providing excitation for the gain medium, the physical spatial overlap between the cavity beam and the pump beam is maximized and most efficient only with the smaller beam size. In this way, the laser cavity operation is "favored" for higher-intensity pulsed operation.

This approach offers a few advantages over passive modelocking with a saturable absorber. First, it is possible to implement without introducing losses and extra physical elements into the cavity (using the soft-aperture scheme). Second, the response of the KLM effect is quasi-instantaneous, bypassing the needed recovery time after bleaching seen in other absorbers. This long recovery time after bleaching can be seen in the asymmetry in the loss curve in Fig. 5(b). This asymmetry also has the undesired effect of still having gain for low-intensity components, temporally broadening the pulse and limiting the pulselength that would be fundamentally achievable.

KLM and pulse generation are typically achieved by introducing noise into the oscillator. This can be done by rapid motion of a mirror or other element, such as a dispersion-compensating prism within the cavity. Such noise, introduced as a "kick" of one of the elements introduces a noise "pulse" within the system that begins the modelocking process. In commercial laser oscillators, such a kick is often introduced by mounting the high reflection end mirror in the oscillator on a piezoelectric actuator that imparts an impulse to the mirror. Under the correct cavity conditions, the noise pulse will extract slightly more energy from the gain medium than the rest of the light in the cavity, thereby initiating the lateral variation of refractive index within the Ti:sapphire crystal. Each time this initial pulse passes through the crystal, it extracts greater energy from the crystal until it dominates the far weaker CW light and a stable pulse is formed. We can observe this process in action. Figure 6 shows the output of a spectrum analyzer monitoring the spectrum of light circulating in the oscillator. The vertical axis is light intensity while the horizontal axis is wavelength, and each grid line corresponds to a 10 nm step. The central vertical axis corresponds to a wavelength of 800 nm. On the left is a trace of CW light (no pulsing) within the cavity. Note that the trace is narrow and sharply peaked with a bandwidth of <1 nm.

After introducing a noise pulse by rapidly moving a prism within the cavity, the bandwidth increases dramatically indicating the collapse of the CW light into a pulse circulating in the cavity (Fig. 6(b)). The full-width-at-half-maximum (FWHM) is $\sim$25 nm corresponding to a pulsewidth of $\sim$40 fs. The relationship between spectral bandwidth and temporal pulse duration is discussed in Section 2.6.

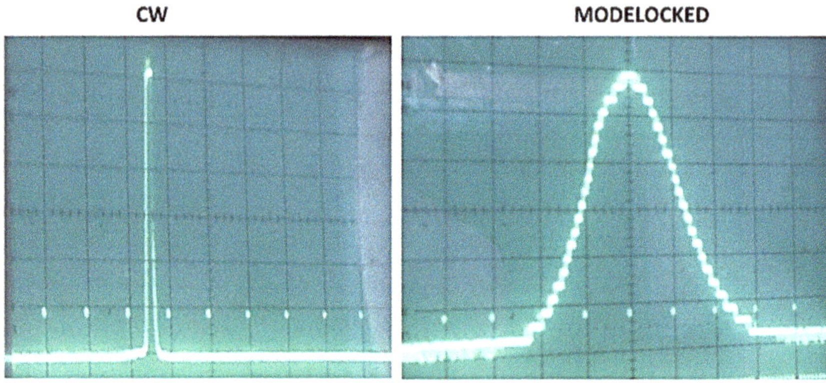

Fig. 6. Spectrum analyzer output from a Ti:sapphire oscillator operating in (a) CW mode and (b) pulsed mode. The spectral bandwidth increases dramatically once a noise pulse is introduced into the cavity. KLM results in the formation of an ultrashort pulse circulating in the cavity — the 25 nm bandwidth corresponds to a ~40 fs pulse in the cavity.

2.3 Ti:Sapphire Oscillators

Ti:sapphire is the most common laser medium used in KLM oscillators. In addition to having the desired nonlinear index of refraction, it has a large gain bandwidth extending from 650 nm to more than 1000 nm and can provide considerable wavelength tunability that allows for the generation of optical pulses down to just a few femtoseconds. This means that short pulses down to the femtosecond regime are possible using Ti:sapphire as compared to narrower bandwidth gain materials that would fundamentally prohibit achieving pulsewidths in the low femtosecond range. This efficient lasing and large absorption cross-section are why Ti:sapphire is almost exclusively used to generate femtosecond light pulses in all applications.

In the crystal itself, sapphire (Al_2O_3) is the host medium, and it is the atomic levels of Ti^{3+} doped within the crystal that are optically excited or "pumped" and will contribute to stimulated emission. The peak in absorption of Ti:sapphire is in the green region of the optical spectrum near 510–540 nm, indicating that Ti:sapphire efficiently absorbs light from commercially available pump lasers such as frequency doubled neodymium-doped yttrium aluminum garnet (Nd:YAlG), yttrium lithium fluoride (Nd:YLF), and yttrium orthovanadate (Nd:YVO_4) that produce light in the 527–532 nm range.

A standard Ti:sapphire oscillator (Figs. 2 and 7) is typically driven by a diode-pumped laser, whose infrared output wavelength near 1,060 nm is frequency doubled by a nonlinear crystal situated within the pump laser

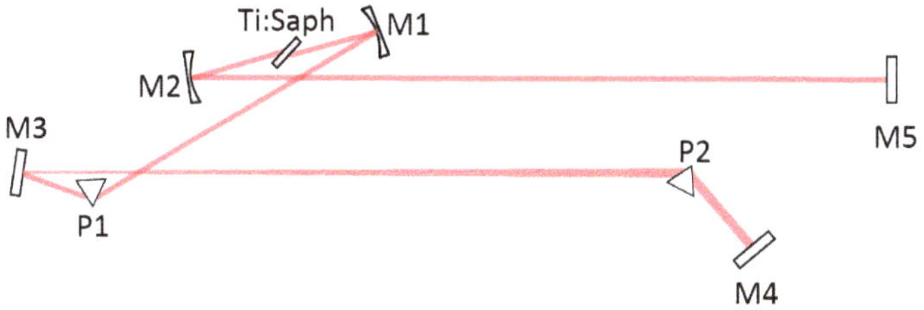

Fig. 7. Typical layout of a dispersion-compensated Ti:sapphire oscillator. M1 and M2 are curved mirrors, M3 is a fold mirror, M4 is an end mirror, M5 is an output coupler, P1 and P2 are fused silica prisms, and Ti:S is the Ti: Sapphire gain medium at Brewsters angle to the incident beam.

cavity, to produce light near 530 nm. This is called intracavity frequency doubling. The output of the intracavity, frequency-doubled pump laser is focused on a Ti:sapphire crystal to create a high gain region within the interior of the crystal. To optimize the cavity for short pulses, dispersion-compensating prisms (prism pair Fig. 7) are located within the cavity, as are mirrors to trap the fluorescent light emerging from the crystal.

One additional important consideration in ultrafast oscillator design is dispersion (Fig. 7). In discussions thus far, the cavity was assumed to be interacting identically with all wavelengths that were propagating within the pulse. However, when light propagates through a material, the refractive index of the material usually has at least some dependence on wavelength. This leads to a difference in phase velocities of the wavelengths, or *dispersion*, in the pulse. In "normal" dispersion materials, longer wavelengths travel faster than shorter wavelengths, giving the pulse a positive frequency sweep or "chirp". For a given input pulselength, τ_0, the output pulselength, τ, can be shown to be

$$\tau = \tau_0\sqrt{1 + \frac{\mathrm{GDD}^2}{\tau_0^4}16(\ln 2)^2}, \tag{5}$$

where GDD is the group delay dispersion due to the gain medium. GDD can be calculated or measured for a given gain medium thickness, typically in units of fs^2. From this relationship, we see that the shorter the pulselength, the more sensitive it is to being broadened by the dispersion due to the gain medium. The Ti:sapphire gain medium is no exception to this, as well as propagation through the air in the cavity. This undesired dispersion can be compensated for by using elements such as negatively chirped mirrors or, more commonly, a set of prisms (P1 and P2) as seen in Fig. 7. In this arrangement, the first prism (P1) acts to angularly disperse the

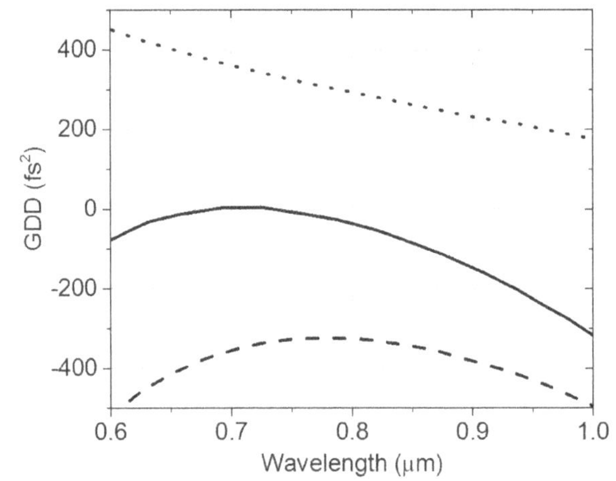

Fig. 8. Calculated group delay dispersion versus wavelength for propagation through a 5 mm long Ti:sapphire crystal (dotted line) and a fused silica prism pair with spacing of 62 cm and 5 mm pathlength through the prisms (dashed line). The resultant GDD is shown with the solid line.

beam, while the second prism (P2) recollimates it. The angular dispersion gives a resultant negative GDD that is proportional to the prism separation distance and can be fine-tuned by adjusting the prism spacing and total pathlength of the beam through the prisms. An example showing the GDD due to the Ti:sapphire gain medium (dotted line) and the negative GDD due to the prism pair (dashed line) is shown in Fig. 8. When combining these elements in the cavity, the net dispersion is then roughly canceled out around the central wavelength of the output broad bandwidth pulse (~750–800 nm). It is important to note that this is not exactly canceled out at all wavelengths, leading to some residual broadening of the pulsewidth. Nonetheless, dispersion-compensation is critical to achieving the shortest pulses.

2.4 Pulse Amplification

A broad range of applications are available utilizing just the fs-range pulses out of a typical oscillator. For example, femtosecond frequency combs have been investigated in both the laboratory and industrial setting to perform precision atomic and molecular spectroscopy (yielding more sensitive chemical sensing devices) and for use in atomic clocks (yielding more precise GPS). However, many applications including industrial machining, femtosecond mask repair, and high-harmonic generation utilized in the spectroscopies

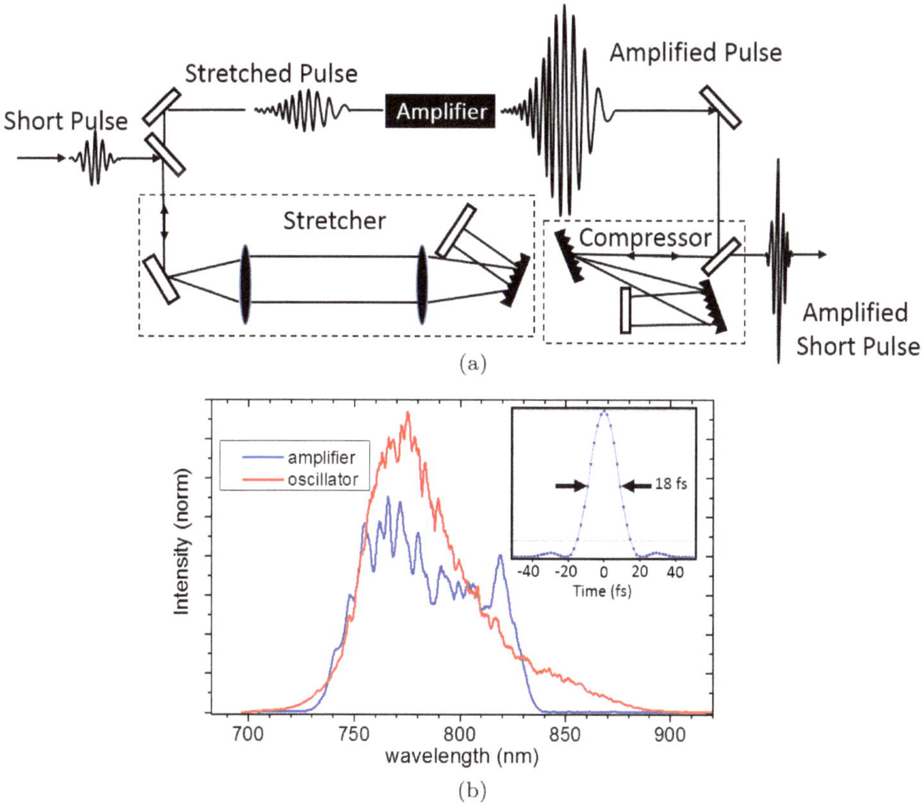

Fig. 9. (a) Chirped Pulse Amplifier layout, adapted from Ref. [5]. (b) Representative oscillator spectrum (red) and resultant spectrum after CPA (blue). Inset shows the transform-limited pulse length of 18 fs from the amplifier bandwidth.

discussed in this book require a higher-intensity pulse than what can be traditionally supplied by an oscillator alone.

The primary method used to accomplish this is Chirped Pulse Amplification (CPA) [17]. A schematic layout of this is shown in Fig. 9. In this method, the initial pulse is stretched in time to hundreds of picoseconds in pulselength to reduce peak power and allow for amplification without damaging optical components or giving nonlinear beam distortions within the amplifying medium. This is an absolute requirement in amplification of laser pulses in crystalline media such as Ti:sapphire because if direct amplification of a femtosecond pulse were attempted, the peak power inside the crystal would quickly exceed the damage threshold for the material. Once the time-stretched pulse is amplified, the pulse can then be recompressed via a set of negative dispersion-compensating gratings. The compressor is designed to reverse the initial stretching of the pulse and to compensate for the extra propagation through the crystal and optics to give an amplified short pulse, usually at a reduced repetition rate in the kHz range. Using this approach, optical pulses have been generated up to the terawatt level.

There are primarily two methods for amplifying the pulse once it has initially been temporally stretched, *regenerative* (Fig. 10(a)) or *multipass* (Fig. 10(b)) *amplification.* In both schemes, it is necessary to decrease the oscillator repetition rate (the number of pulses/second) entering into the amplifier to allow sufficient build up of energy within the amplifier crystal. This can be done with a Pockels cell, consisting of an electro-optic crystal and electrodes to modulate the field applied across the crystal. Using an applied voltage to change the phase delay in the crystal, we are able to

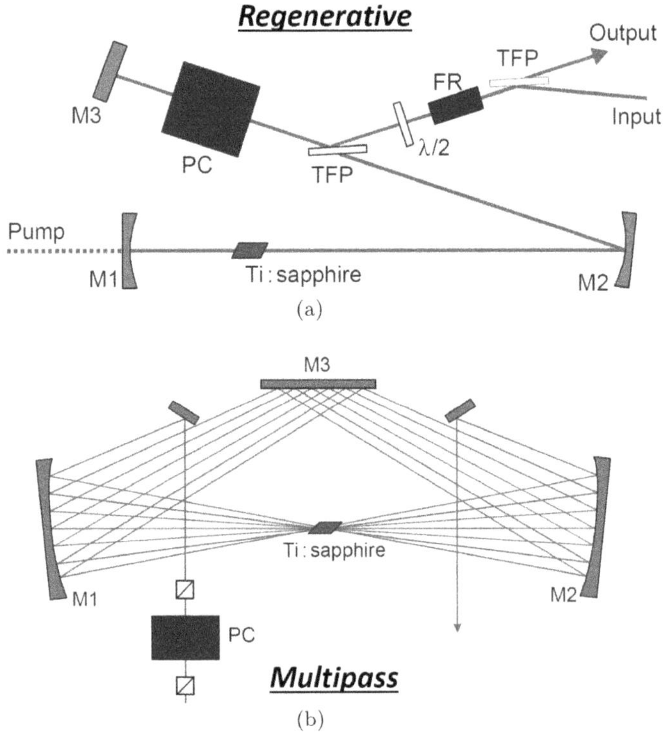

Fig. 10. Simplified optical element layouts for (a) Regenerative amplifier — comprising of a thin-film polarizer (TFP), thin film polarizing beam splitter, Pockels cell, Faraday rotator (FR), and half-wave plate. M1, M2 are curved mirrors, and M3 is a flat mirror. (b) Multipass amplifier — M1, M2 are curved mirrors, M3 is a flat mirror, and PC is the Pockels cell that is bracketed by a pair of polarizers for single pulse injection. Adapted from Ref. [19].

create a "voltage-controlled waveplate", where a desired repetition rate can be chosen to have a unique polarization relative to the rest of the pulse train. Subsequently using an appropriate waveplate, the rest of the pulses can then be discarded.

2.4.1 *Regenerative Amplifier*

Having reviewed the principles of active modelocking, the overall implementation scheme in a regenerative amplifier will seem familiar. In active modelocking, a physical element like an electro-optical modulator was inserted into the cavity and periods of low loss were synchronized with the resonator round-trip time with the goal to produce a pulsed output. In a regenerative amplifier, a similar electro-optical modulator is inserted into the cavity but with the goal to limit pulse propagation in the cavity in order to sufficiently amplify the temporally stretched pulse. In this scheme (shown in Fig. 10(a)), the amplifier gain medium is initially allowed to accumulate energy by being pumped by a secondary laser. In the case of a Ti:sapphire gain medium, this is similarly done via 527–532 nm pump laser line in the Ti:sapphire oscillator case. The combination of an FR, TFP, and half-wave plate are used to separate the input and outgoing amplified pulse train in the cavity.

The process of amplifying a nanojoule pulse from the oscillator in a regenerative amplifier involves several steps. The first step involves the realization that a femtosecond laser at 800 nm pulse would severely damage the Ti:sapphire crystal as it nears its peak amplified value. Hence, prior to injecting the pulse into the regen cavity, it must be stretched significantly from the femtosecond range to many nanoseconds so that the peak intensity of the pulse is below the damage threshold of the crystal. Four bounces from an appropriate grating are required before injection into the regen.

In a common regenerative amplifier integration scheme (Fig. 10(a)), a single Pockel cell triggered at the appropriate time, and working in concert with a TFP and $^1/_4$-wave plate, acts to "trap" one of the injected pulses within the cavity. Light from the oscillator is typically vertically polarized (S-polarization) so that it efficiently reflects from the TFP oriented at Brewster's angle into the regen cavity. Once in the cavity, the Pockels cell/waveplate tandem rotates the polarization of the pulse to horizontal (P-polarization) so that the pulse passes through the TFP upon its return. The pulse is now trapped in the cavity where, through multiple round trips through the Ti:sapphire gain medium, it is amplified by $\sim 10^6$ (from nJ to mJ pulse energy). To eject the now amplified pulse, the Pockels cell is triggered and the pulse now rotates back to S-polarization and is reflected out of the regen cavity.

The amplified pulse possesses a good deal of energy, but must be compressed back to its near-original pulsewidth. This is done in the compressor section of the system where the pulse again undergoes four diffractive reflections from a grating, essentially the reverse of the original stretching process.

High gain can be achieved with this approach, giving desirable high-pulse energies while maintaining kHz repetition rates. This is especially helpful in applications to micromachining since higher rep rate and high-pulse energies mean accurate ablation, shorter exposure needed, and faster overall machining time. However, there are several limitations to this scheme. Primarily, the limited gain bandwidth of the amplifier medium means the propagating pulse bandwidth is also limited, broadening the pulselength. With typical configurations, regenerative amplifiers are usually limited to ~40 fs pulselength. Figure 11 shows pictures of an oscillator/regenerative amplifier pair in the lab.

2.4.2 *Multipass Amplifer*

Instead of relying on tuned electro-optic elements to isolate the propagating pulses within the cavity, multipass amplifiers take a simpler and, arguably, more "brute force" approach. In this design, a set number of passes are aligned through the gain medium geometrically, as shown in Fig. 10(b). An initial steering mirror redirects the beam onto a curved mirror (M1), focusing the beam into the pumped gain medium. A second curved mirror (M2) collimates the beam and redirects it to a flat, folding mirror (M3). This configuration is aligned such that each pass around the ring is slightly offset from the pass before, allowing the beam to be kicked out of the "ring" after a set number of passes. This type of arrangement is able to amplify input power by 10^6–10^8 times and can amplify nJ pulses to mJ levels. At low rep rates (10 Hz), even multi-Joule pulses have been reported [18, 19].

An important detail to note is the geometric constraint that the Ti:sapphire crystal remains relatively thin to provide a suitable "window" to pass the propagated beam through at increasingly extreme angles (see Fig. 10(b)). This gives the benefit of both increased broadband amplification compared to the regenerative amplifier case and limited nonlinear dispersion due to the crystal. As a consequence, the outgoing pulses in multipass

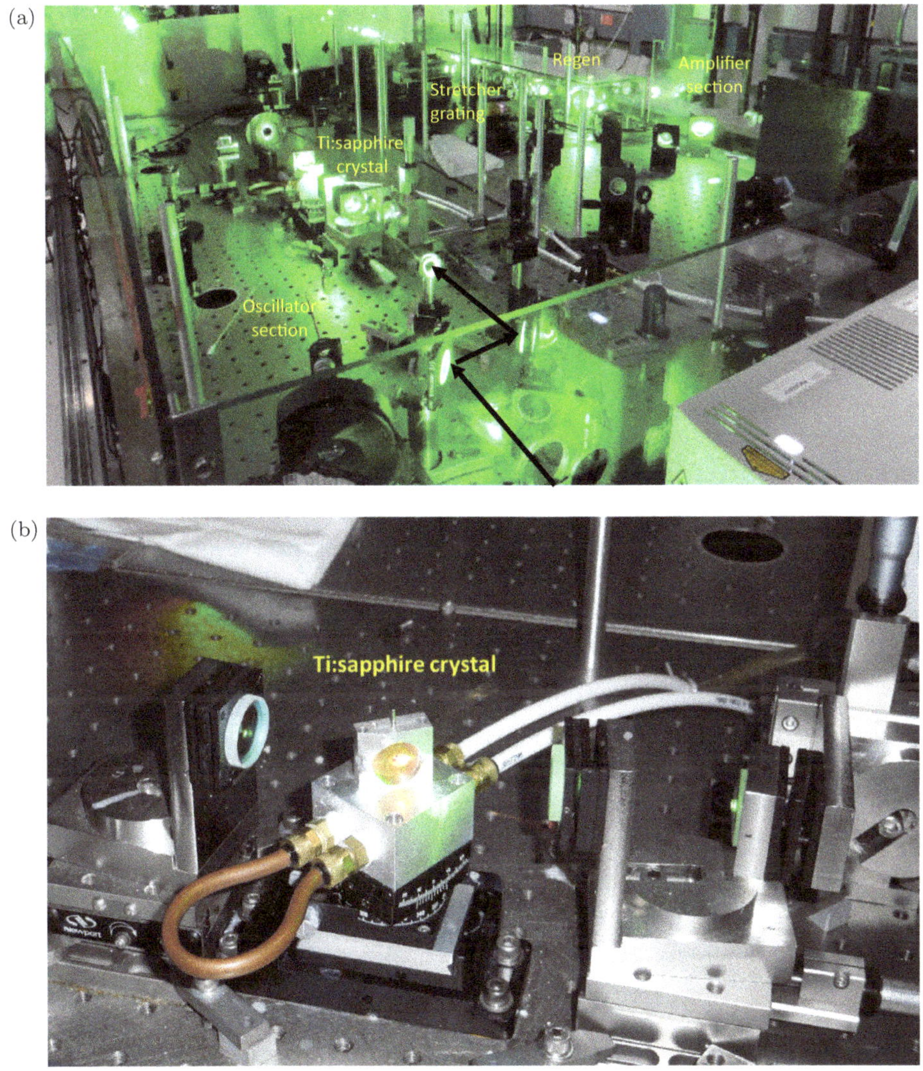

Fig. 11. (a) Layout of oscillator and amplifier. Left side of picture is the oscillator pumped by an intracavity doubled Nd:YLF laser near bottom of picture. Output from the oscillator is directed to the stretcher grating where through four passes the pulse length is stretched to the nanosecond timescale and injected into the regen.

amplifier systems are able to be compressed to even shorter pulselengths (<25 fs). The primary drawback with this scheme is layout complexity. Considerable expertise is required to align this amplifier and to maintain high gain; as a result, multipass amplifier systems are still mainly limited to lab environments.

2.4.3 *Emerging Oscillator/Amplifier Designs*

While the majority of oscillators and amplifiers used in industrial applications are variants of the Ti:sapphire design discussed, there are several designs that are rapidly making progress in yielding highly stable, compact designs with pulselengths <200 fs. In particular, compact fiber laser systems have already been applied in the areas of micromachining [20], communications [21], trace gas sensing [22], and medical imaging [23]. There are trade-offs between performance and compactness, but for industrial applications the latter is quite important as these systems are often incorporated into larger tools and manufacturing floor space always comes at a premium.

The solid-state laser systems described earlier suffer from a few drawbacks. First, direct diode pumping of the gain medium, which would give a higher efficiency process, is not possible. Second, the thickness of the gain medium itself leads to trade-offs in output power and performance. Higher single-pass gain and therefore, higher output powers, result from a thick gain medium at the expense of introducing thermal lensing effects. Conversely, a thinner gain medium overcomes these thermal effects, but limit the system's power scaling capabilities [24, 25]. Both of these concerns are minimized in ultrafast fiber lasers. These systems usually incorporate a long, polarization maintaining, optical fiber, typically SiO_2 doped with erbium, neodymium, or ytterbium as its gain medium. As such, thermal dissipation is maximized due to the large ratio of surface-to-active volume of the fiber. Power-dependent distortions are also minimized as the geometry of the fiber maintains high beam quality. With regard to efficiency, the spatial confinement of the laser and pump beam within the optical fiber greatly extends their interaction length. In Ti:sapphire lasers, this interaction length is limited to a small volume near focus of the Gaussian beam; this is the *Rayleigh length*, defined as the distance between focus and the where the beam doubles its minimum size. This serves to greatly increase the output efficiencies (above 80% for ytterbium doped) and decrease residual heating. Additionally, double-clad fibers have become more commonly employed, whereby a multimode cladding surrounds the inner fiber core and serves to gradually convert any remaining lower power pump light into laser radiation. This further increases the conversion efficiency of the process.

With all of these advantages, several drawbacks exist that have limited the ability to achieve pulselengths in these systems to as short as those seen in Ti:sapphire systems, for a given output power. In particular, these

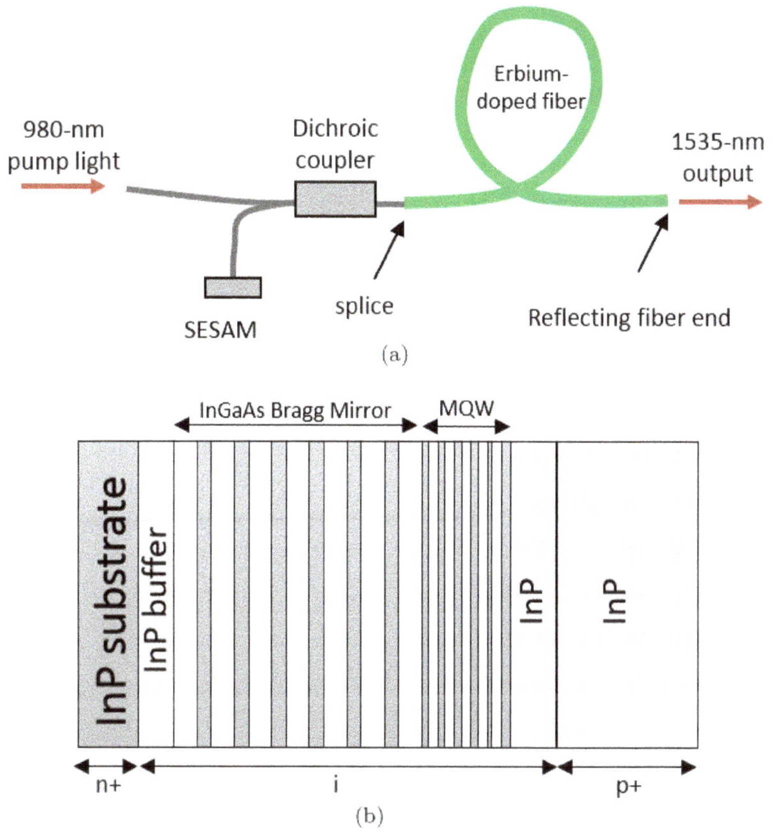

Fig. 12. Simple erbium-doped femtosecond laser design, modeled after Ref. [26] showing (a) general overview of the laser and (b) SESAM composition for the case of Ref. [26]. n+, I, and p+ denote the doping for each region of the multilayer stack.

systems suffer from nonlinear effects due to the long propagation lengths within the fiber. The pulselengths from fiber systems are typically found at or just below 200 fs. Methods for dispersion compensation have, therefore, become a key field of interest in developing robust, short pulse systems (for a general overview, see Ref. [26]).

A simple implementation of an erbium-doped femtosecond laser is shown in Fig. 12. The laser cavity is defined on one end by a reflecting fiber end (acting as the output coupler seen in the Ti:sapphire design) and a semiconductor saturable absorber mirror (SESAM) on the other. The SESAM serves to passively modelock the cavity and consists of a Bragg mirror and a multiple quantum well (MQW) absorber layer. The thickness and compositions of the Bragg mirror and MQW multilayers can be optimized for the desired laser wavelength.

The doped optical fiber (Er, in the case shown in Fig. 12) is pumped using a laser diode and inserted into the cavity via a dichroic fiber coupler. Unlike the Ti:sapphire case, where the Pockels Cell was needed to determine repetition rate of the output pulses, both the pulse length and repetition rate can be controlled here by tuning the total length of the fiber. This is dictated by the specific balancing of dispersion within the fiber, gain within the medium, and nonlinear effects. For the interested reader, several excellent reviews exist to discuss the finer points of alternative designs [27].

2.5 Time-Bandwidth Product

A key to understanding the relationship between the frequency bandwidth of a pulse and its temporal width can be seen in the time and frequency transforms of a pulse. If we start with the premise that a pulse can be represented by a Gaussian function, and the Fourier transform of a Gaussian function is also Gaussian, then the time and frequency transforms can be written as

$$E(t) = \frac{1}{2\pi} \int_{-\infty}^{\infty} E(\omega) e^{-i\omega t} d\omega \tag{6}$$

and

$$E(\omega) = \frac{1}{2\pi} \int_{-\infty}^{\infty} E(t) e^{i\omega t} dt. \tag{7}$$

Here, $E(\omega)$ and $E(t)$ are the frequency, and time-dependent electric fields of the pulse itself. As will be described, the temporal width of a pulse is measured in a device called an autocorrelator. In particular, we will discuss the most convenient of these devices, called a single-shot autocorrelator.

The frequency bandwidth is also an important quantity to be measured. This is measured in a device called a spectrum analyzer. For both the pulse temporal width and frequency bandwidth, an experimentally convenient quantity to be measured is the FWHM.

The two expressions (6) and (7) are related by the inequality below, known as the time-bandwidth product

$$\Delta t \Delta \nu \geq K. \tag{8}$$

Here, $\omega = 2\pi\nu$ and $\Delta\nu$ is the frequency bandwidth that is measured as the FWHM in the spectrum analyzer, and Δt is the FWHM measured in an autocorrelator. K is a number that depends on the particular pulse shape. For femtosecond pulses generated from an oscillator or amplifier, the pulse shape is typically Gaussian $\left(e^{-\frac{1}{2}\left(\frac{t}{to}\right)^2}\right)$ or hyperbolic secant $(\sec h\left(\frac{t}{t_o}\right))$. For the Gaussian case $K = 0.441$ and for hyperbolic secant $K = 0.315$.

As an example, consider an 800 nm pulse whose frequency bandwidth is 30 nm. Since $\nu = c/\lambda$, then

$$\Delta t \geq \frac{K\lambda_o^2}{c\Delta\lambda}, \tag{9}$$

where λ_o is the central frequency (800 nm) and $\Delta\lambda$ is the wavelength bandwidth. For our 30 nm bandwidth and a Gaussian pulse where $K = 0.441$, the calculation gives a lower limit to the pulsewidth of 31 fs. For this case, the pulse is said to be "transform-limited". This calculation is important since, as we mentioned, a standard spectrum analyzer provides the nominal spectral bandwidth of the pulse. If it is important to achieve a sub-50 fs pulse, then it is critical to adjust the dispersion in the oscillator used to generate these pulses to achieve a sufficient bandwidth to provide for the required short pulses.

2.6 Single-Shot Autocorrelation

Measuring the width of a femtosecond pulse is a challenge. There are no photodiode detectors with a temporal response sufficiently fast to capture and measure a pulse of light with such a short duration. In order to measure the time width of such a pulse, nonlinear optical techniques must be employed.

Since the polarization of a dielectric is proportional to the applied electric field, we can write

$$P = \varepsilon_o \chi E. \tag{10}$$

But for the general case where more intense electric fields must be considered, such as those found in a femtosecond pulse of light, the polarization can be expressed as a Taylor expansion

$$P(t) = \varepsilon_o(\chi^1 E(t) + \chi^2 E(t)^2 + \cdots). \tag{11}$$

Here, $P(t)$ is the polarization of the dielectric that varies with the strength of the oscillatory electric field, $E(t)$, the applied electric field and χ is the electric susceptibility. For increasingly intense pulses, the electric field is, in turn, stronger, and so higher orders of the susceptibility in the expansion must be considered.

The first step is to exploit the frequency doubling properties of non-centrosymmetric (lacking a center of inversion) crystals with the appropriate optical properties. Crystals such as potassium dideuterium phosphate (KDP) and beta-barium borate (BBO) are among such media that possess a strong nonlinear response when intense pulses of light transit through the crystal. There are a number of ways to generate the second harmonic of a pulse of light. By second harmonic we mean that for an input light pulse with a central wavelength of 800 nm, the second harmonic generated would then have a central wavelength of 400 nm.

For a sufficient intensity of second harmonic light to be observable emerging from the crystal, the output of the second harmonic light must be in phase with the incident laser light; this is a condition called phase matching. Consider a light pulse of frequency ω incident upon the crystal. The input electric field at frequency ω is of the form

$$E_1(t) = E_1 \cos(\omega t - k_1 r), \tag{12}$$

while the output electric field at 2ω is

$$E_2(t) = E_2 \cos(2\omega t - k_2 r). \tag{13}$$

The second-order polarization producing the second harmonic is

$$P^{(2)}(t) = \varepsilon_o \chi^{(2)} E^2 \cos(2(\omega t - k_1 r)), \tag{14}$$

where $k_2 = 2k_1$. Note also that the polarization depends on the *square of the electric field*. This means that the second harmonic is dependent upon the *intensity* of the input laser light and the pulse observed in the autocorrelator described below will show a width that is dependent upon intensity and must be corrected for the pulse type to give the proper pulsewidth. For a Gaussian pulse shape, the intensity is

$$I(t) \sim e^{\frac{-4ln2t^2}{\Delta t^2}} \quad \frac{\Delta t}{t} = \sqrt{2} \tag{15}$$

and for a hyperbolic secant pulse

$$I(t) \sim \text{sech}^2\left(\frac{1.76t}{\Delta t}\right) \quad \frac{\Delta t}{t} = 1.55, \tag{16}$$

indicating that for a Gaussian pulse one needs to divide the width of the autocorrelation trace by 1.414 and a hyperbolic secant pulse by 1.55 to get the true pulse width.

Figure 13 shows a schematic for a single-shot autocorrelator. A single pulse is split into two correlated beams and then recombined in a second harmonic crystal such as KDP or BBO. By symmetry, and for the correct

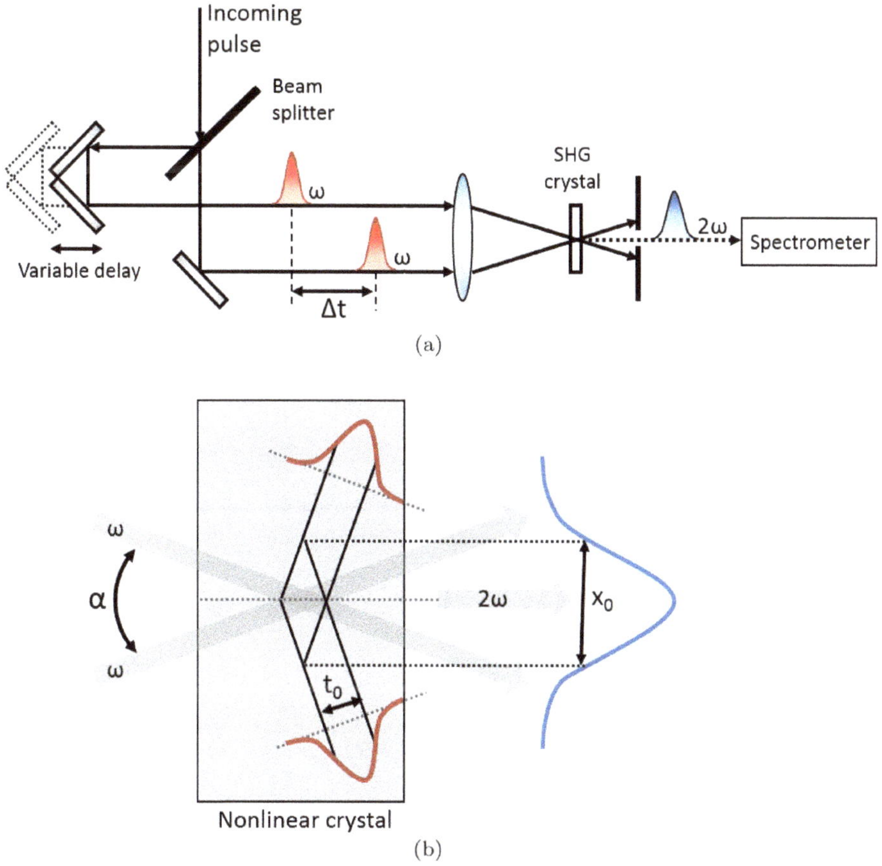

Fig. 13. (a) schematic layout of a single-shot autocorrelator. A single light pulse is split into two and recombined within a nonlinear crystal such as KDP. When the pulses are both spatially and temporally overlapped, the second harmonic light (of frequency 2*omega) emerges and is measured by a linear charge coupled device array. (b) View of pulse interaction within the nonlinear crystal, with beams incident at an angle (alpha).

spatial and temporal overlap of the pulses within the crystal, the second harmonic is generated and directed toward the detector. The detector consists of a one-dimensional array of charge-coupled devices that measure the intensity of the frequency doubled light along the array itself. The bottom schematic shows the detailed overlap of the two pulses within the nonlinear crystal.

Figure 14 top, shows a picture of an actual single-shot autocorrelator. In Fig. 14(b), the KDP crystal is the white unit at the bottom where the two light pulses converge. The small box at the bottom left in the picture controls the delay line where the stretched pulse is recompressed following amplification, as described in Section 2.4.

(a)

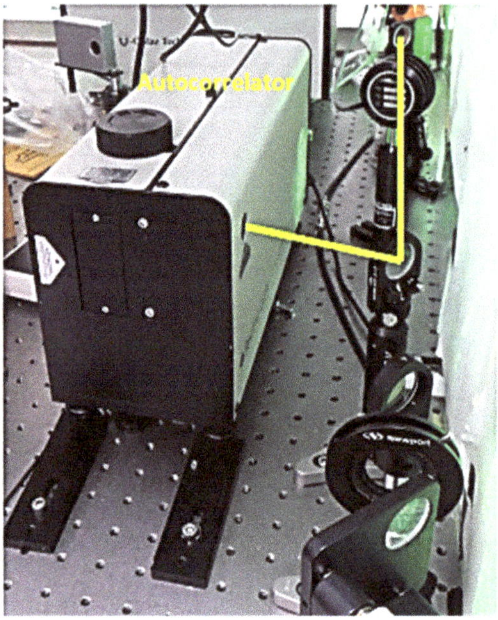

(b)

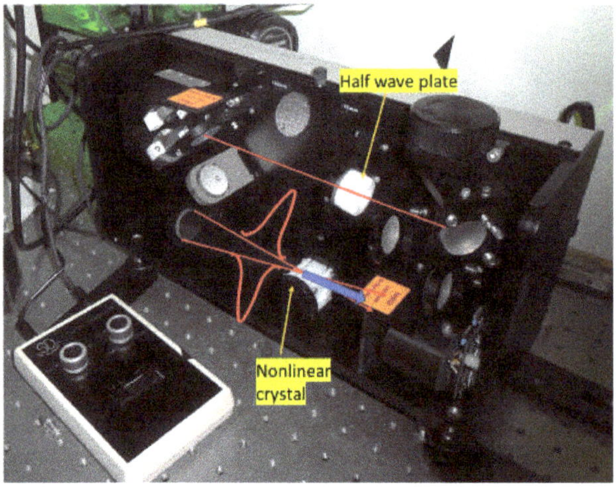

Fig. 14. (a) single-shot autocorrelator. (b) single-shot autocorrelator internal optics.

The output of the single-shot autocorrelator is displayed in Fig. 15 as an analog trace on an oscilloscope. The FWHM is ~50 fs and is close to the 40 fs oscillator pulse that is injected into the regenerative amplifier. The stepped granularity seen in Fig. 15 is associated with the individual charge-coupled devices as the width of the measured amplified pulse is nearing the time resolution of the autocorrelator.

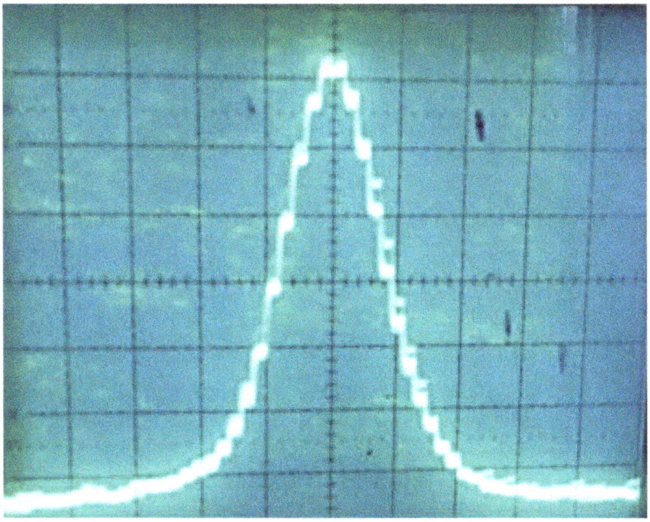

Fig. 15. Output trace from the single-shot autocorrelator. Each grid line corresponds to 26 fs

2.7 Summary

This chapter was devoted to a somewhat practical discussion of the operation and function of ultrafast laser systems, including oscillators and amplifiers, for use in industrial settings. The key considerations involve coupling an oscillator, used to create the short pulse with an amplifier that increases the pulse energy by more than 10^6 to useful levels. Ti:sapphire-based laser systems are by far the most heavily used for this purpose in both academic and industrial settings since they are robust, reliable, and stable over extended periods of time. With a few key metrological capabilities, including a spectrum analyzer to monitor the bandwidth of pulses circulating in the oscillator and a single-shot autocorrelator, sufficient characterization of femtosecond pulses can be obtained.

References

[1] A. E. Siegman, *Lasers.* University Science Books, South Orange, 1986.

[2] W. T. Silfvast, *Laser Fundamentals*, Cambridge University Press, Cambridge, 2004.

[3] H. A. Haus, J. G. Fujimoto, and E. P. Ippen, *IEEE J. Quantum Electron.*, vol. 28, p. 2086, 1992.

[4] T. Brabec, C. Spielmann, P. F. Curley, and F. Krausz, *Opt. Lett.*, vol. 17, 1292, 1992.

[5] G. Cerullo, S. De Silvestri, and V. Magni, *Opt. Lett.*, vol. 19, p. 1040, 1994.

[6] P. Albers, E. Stark, and G. Huber, *JOSA B*, vol. 3, p. 134, 1986.

[7] P. M. W. French, J. A. R. Williams, and J. R. Taylor, *Opt. Lett.*, vol. 14, p. 686, 1989.
[8] D. J. Kuizenga, D. W. Phillion, T. Lund, and A. E. Siegman, *Opt. Commun.*, vol. 9, p. 221, 1973.
[9] P. P. Sorokin, J. J. Luzzi, J. R. Lankard, and G. D. Pettit, *IBM J. Res. Dev.*, vol. 8, p. 182, 1964.
[10] F. J. McClung and R. W. Hellwarth, *J. Appl. Phys.*, vol. 33, p. 828, 1962.
[11] A. J. DeMaria, W. H. Glenn, M. Brienza, and M. E. Mack, *Proc. IEEE*, vol. 57, p. 2, 1969.
[12] J. Zhou, I. P. Christov, G. Taft, C.-P. Huang, M. M. Murnane, and H. C. Kapteyn, *Opt. Lett.*, vol. 19, p. 1149, 1994.
[13] J. A. Valdmanis, R. L. Fork, and J. P. Gordon, *Opt. Lett.*, vol. 10, p. 131, 1985.
[14] U. Keller, *Nature*, vol. 424, p. 831, 2003.
[15] D. H. Sutter, L. Gallmann, N. Matuschek, F. Morier-Genoud, V. Scheuer, G. Angelow, T. Tschudi, G. Steinmeyer, and U. Keller, *Appl. Phys. B Lasers Opt.*, vol. 70, p. S5, 2000.
[16] T. Brabec, P. F. Curley, C. Spielmann, E. Wintner, and A. J. Schmidt, *JOSA B*, vol. 10, p. 1029, 1993.
[17] D. Strickland and G. Mourou, *Opt. Commun.*, vol. 55, p. 447, 1985.
[18] C. Le Blanc, G. Grillon, J. P. Chambaret, A. Migus, and A. Antonetti, *Opt. Lett.*, vol. 18, p. 140, 1993.
[19] S. Backus, J. Peatross, C. P. Huang, M. M. Murnane, and H. C. Kapteyn, *Opt. Lett.*, vol. 20, p. 2000, 1995.
[20] S. Lawrence and M. E. Fermann, (2006).
[21] H. Ohta, S. Nogiwa, N. Oda, and H. Chiba, *Electron. Lett.*, vol. 33, p. 2142, 1997.
[22] I. Galli, S. Bartalini, S. Borri, P. Cancio, D. Mazzotti, P. De Natale, and G. Giusfredi, *Phys. Rev. Lett.*, vol. 107, 2011.
[23] T. Juhasz, G. Kastis, C. Suarez, Z. Bor, and W. Bron, *Lasers Surg. Med.*, vol. 19, p. 23, 1996.
[24] M. E. Fermann and I. Hartl, *IEEE J. Sel. Top. Quantum Electron.*, vol. 15, p. 191, 2009.
[25] M. Gebhardt, C. Gaida, F. Stutzki, S. Hädrich, C. Jauregui, J. Limpert, and A. Tünnermann, *Opt. Lett.*, vol. 42, p. 747, 2017.
[26] W. Loh, D. Atkinson, P. Morkel, M. Hopkinson, A. Rivers, A. Seeds, and D. Payne, *Appl. Phys. Lett.*, vol. 63, 1992.
[27] M. Fermann and I. Hartl, *Nat. Photonics*, vol. 7, p. 868, 2013.

Chapter 3

High-Harmonic Generation with Femtosecond Light

3.1 Introduction

Development of bright Extreme Ultraviolet (EUV) sources has been of great interest to industry, with the primary application being EUV lithography to enable the printing of ever-smaller features for semiconductor devices. However, laser-like EUV sources with high spatial and temporal coherence offer a wide array of relevant applications desired for material characterization (via techniques like photoemission spectroscopy and thermal transport measurements), imaging of nanoscale features, and basic molecular and atomic spectroscopy. Generating this laser-like light has proved difficult to realize due to the tendency for most materials to be highly absorbing at shorter wavelengths than 200 nm. This greatly limits the choice of materials that can be used as a laser gain medium and makes production via traditional nonlinear crystals infeasible. In general, the pump power required to produce a given output wavelength via a traditional laser roughly scales as λ^{-5} [1, 2]. This would mean an output wavelength in the X-ray region of 1 nm would already require terawatt pump power. Due to this difficulty, EUV/soft X-ray lasers were not demonstrated until 1994 [3]. Using EUV lasing transitions in gases and fast capillary discharges as the lasing medium, these table-scale lasers have output wavelengths ranging from 46.9 nm to less than 8 nm, but with only picosecond pulselengths or longer.

Since its first observation in 1987 [4], high-harmonic generation (HHG) has been one of the most viable candidates for a versatile, compact source of EUV to soft X-ray light with femtosecond to attosecond pulselengths. In this process, coherent visible light is nonlinearly upconverted to EUV and soft X-ray wavelengths, generating a comb of harmonics that ideally spans the entire spectrum between the fundamental and the EUV/X-ray region. As will be described in further detail, HHG involves the interaction of a very intense femtosecond pulse of light, typically in the near-IR (800 nm and longer) with a noble gas such as He, Ne, Ar, Kr, or Xe. At intensities of $\sim 10^{14}$W/cm^2 or higher the atoms within the light field of the pulse are ionized through field-induced tunneling. The emitted electrons are not really free but instead are captured by the optical field of the intense laser pulse and, under the right conditions, a small fraction are driven back toward the nucleus. Recombination is then possible and results in the emission of radiation. Since the process is driven in phase with the optical pulse, the emitted radiation is coherent.

The extraordinarily high intensities needed to drive HHG required the invention and development of the table top laser sources as described in Chapter 2. These ultrafast, laser systems could reach intensities as high as 10^{15}–10^{18} W/cm^2 by combining ultrashort pulse lengths, high-pulse energies, and appropriate focusing conditions. Recent work in improved generation regimes with long driving wavelengths (>1 μm) have pushed photon energies up to the keV regime [5, 6]. Isolated attosecond pulses have also been generated using similar near and mid-IR driving lasers [7, 8].

Historically, various laser sources, in addition to Ti:sapphire-based systems, were shown to be useful in producing high harmonics, including amplified dye laser [9–11] and excimer-based systems [12]. The flexibility associated with different driving laser sources for converting the light into the EUV region has several advantages. Due to HHG having its characteristics dependent upon the driving laser source, femtosecond and even attosecond pulses can be generated robustly using femtosecond laser sources like the Ti:sapphire oscillator/amplifier systems. Additionally, the ability to pick from a range of generated harmonic wavelengths also allows for a wider range of material applications. This combination of high spatial and temporal coherence with favorable energy and temporal resolution on a continuously accessible tabletop source has become an attractive technique used increasingly in laboratory and industrial settings.

3.2 Semi-Classical Model

While a quantum mechanical description is needed to obtain an accurate model of HHG [13, 14], approximations can be made that give a more intuitive semi-classical, quasi-static "three step model". This model was first proposed by Kulander [15] and Corkum [13]. Initially, a femtosecond laser pulse is focused into the medium. This is typically a gas, as in the presented case, but clusters [16], molecules [17], and solids [18] have also been used. Through ionization of the individual atoms, a co-propagating beam of the high harmonic and fundamental beam is created. Multilayer optics, spectral

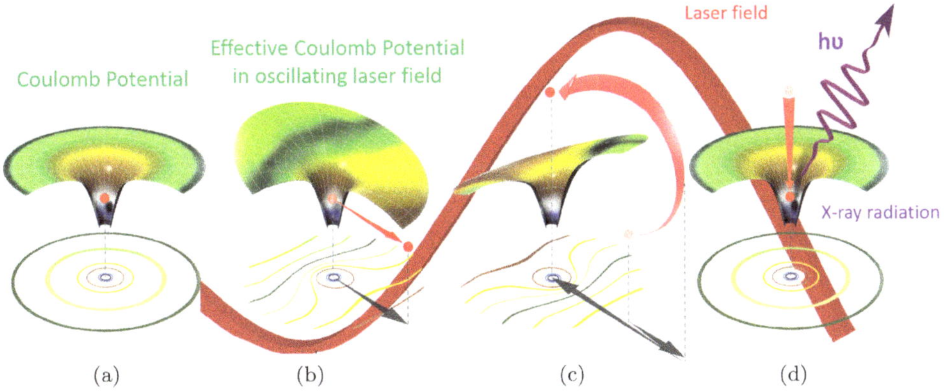

Fig. 1. Three Step model of HHG where (a) the ground state is initially perturbed by the driving laser field to ionize the atom (b–c) the field accelerates and returns the electron back to the parent atom, and (d) results in the recombination and relaxation of the electron to the ground state. This emits a photon that can span up into EUV/soft X-ray wavelengths. Adapted from Ref. [7].

gratings, and filters, can then be used to isolate specific wavelengths of the generated harmonics for the particular application.

Upon focusing the laser pulse into the gas medium, the three steps of Kulander and Corkum's theory (a conceptual illustration of which is shown in Fig. 1) can be modeled as follows:

- **Ionization** of an electron from the atom by an intense laser field.
- **Motion in the field after ionization:** Acceleration of the free electron in the laser field, followed by return to the ion upon the laser electric field switching directions.
- **Recombination of the free electron with the parent ion:** Emitting a photon with energy dependent on the ionization potential of the parent atom and the classical kinetic energy gained by its acceleration in the field.

With each step having an associated probability of occurrence, the total probability of occurrence per atom can be calculated to predict the output intensity of each harmonic order. The details of each of these processes will be discussed individually in the following sections.

3.3 Ionization

A Coulomb potential is used to model the initial, unperturbed atomic potential. Depending on the strength of an external electric field from the laser pulse, this potential can be weakly to strongly distorted, resulting in three possible ionization regimes of the valence electron: multiphoton ionization,

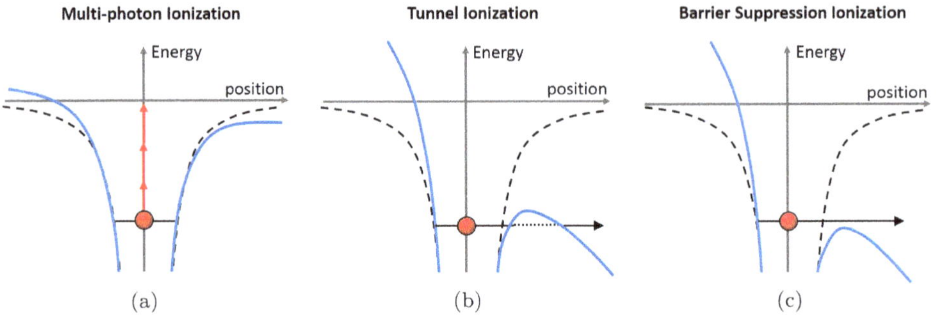

Fig. 2. The three possible ionization potential schemes are (a) multiphoton ionization (b) tunnel ionization (c) barrier suppression ionization. The dashed curve represents the unperturbed Coulomb potential with the blue curve being the effective potential when including the driving laser field. The horizontal and vertical gray axes represent the position and binding energy, respectively (adapted from Ref. [1]).

tunnel ionization, and "barrier suppression" ionization [13, 19]. An illustration of these regimes is shown in Fig. 2. The factor that is commonly used to determine which of these processes is dominant is the *Keldysh parameter* which relates the incident laser frequency to the tunneling frequency

$$\gamma = \frac{\omega_{\text{laser}}}{\omega_{\text{tunnel}}} = \sqrt{\frac{I_p}{2U_p}}, \tag{1}$$

where I_p is the ionization potential of the atom and U_p is the pondermotive potential — the "classical" kinetic energy gained by the electron accelerating in the laser electric field. The pondermotive potential is defined as

$$U_p = \frac{e^2 E_0^2}{4m\omega^2}, \tag{2}$$

or directly in terms of the laser intensity (W/cm^2) and driving wavelength (in μm)

$$U_p \approx 9.33 * 10^{-14} I\lambda^2, \tag{3}$$

where e is the electron charge, E_0 is the electric field of the driving laser, I is the laser intensity, m is the mass of the electron, and ω is the angular frequency of the driving field. Consider the following three criteria for the magnitude of γ, the Keldysh parameter:

- $\boldsymbol{\Upsilon \gg 1}$**:** If the laser intensity is small ($\Upsilon \gg 1$), giving little perturbation to the initial Coulomb potential, then multiphoton ionization is dominant (Fig. 2(a)). Multiphoton ionization is the case where multiple photons that are coherent, in-phase, and present within a short period of time (femtosecond pulse) cumulatively contribute to excite an atomic electron above the vacuum level (ionization energy) of the atom. As such, the resultant ionization can be treated with perturbation theory. Since the electron does not have enough time to tunnel through the potential barrier during each laser cycle, it is able to "bounce" back and forth within the potential as the electric field oscillates until enough photons are absorbed to ionize.

- $\boldsymbol{\Upsilon < 1}$: If $\Upsilon < 1$, then the Coulomb potential is severely distorted by the electric field and the electron can undergo tunnel ionization through the suppressed effective potential barrier, shown with the dotted line in Fig. 2(b). The rate at which it ionizes has been described by Ammosov *et al.* [20]. For the case of argon ($I_p = 15.76$ eV) illuminated with an 800 nm pulse, this regime is dominant for $I_p > 10^{14}$ W/cm^2. With laser intensities in the range 10^{14}–10^{16} W/cm^2, this ionization process is dominant in most high harmonic generation schemes.
- $\boldsymbol{\Upsilon \ll 1}$: Finally, for $\Upsilon \ll 1$, the population of electrons in the ground state can be easily ionized since the effective potential is suppressed below the ionization barrier (Fig. 2(c)). This behavior is called barrier-suppression ionization.

3.4 Acceleration

Once the electron is ionized, the laser field intensity is much greater than the Coulomb potential and it can be modeled as free to evolve in the field. Since the electron effectively has a continuum of states available to it, it can be modeled classically instead of strictly quantum mechanically. The electric field therefore has the form

$$E(t) = E_0 \cos(\omega t)\hat{e}_x + \alpha E_0 \sin(\omega t)\hat{e}_y, \tag{4}$$

where α represents the polarization of the fundamental light (0 for linear and ± 1 for circularly), E_0 is the amplitude of the electric field, ω is the field frequency, and $\hat{e}$ is the unit vector denoted in the x and y directions. Since the electron needs to eventually recombine with its parent ion, this can only happen with $\alpha = 0$, requiring our incident light to be linearly polarized when using one driving field. The recent realization of circular harmonics [21, 22], can be achieved through the use of multiple driving fields. The two-dimensional equations of motion for the electron can be obtained through solving the familiar force equation of $ma = eE$ as

$$v_x(t) = \frac{eE_0}{\omega m}\sin(\omega t) + v_{0x}, \tag{5}$$

$$x(t) = -\frac{eE_0}{\omega^2 m}\cos(\omega t) + v_{0x}t + x_0, \tag{6}$$

$$v_y(t) = -\frac{\alpha e E_0}{\omega m}\cos(\omega t) + v_{0y}, \tag{7}$$

$$y(t) = -\frac{\alpha e E_0}{\omega^2 m}\sin(\omega t) + v_{0y}t + y_0. \tag{8}$$

This assumes that the electron is at rest and at the origin once it tunnels through the potential barrier. This is a reasonable assumption since the displacement position after tunneling is small compared to its maximum deviation position from the atom (Å versus nm). Only electrons released within a specific range of driving laser phases (ϕ) will allow the electron to actually return to the parent atom. Plotting a range of electron trajectories depending on the tunneling phase (shown in Fig. 3(a)), one can see this only occurs between 0° and 90°.

The resultant final kinetic energy of the electron accelerating in the field depends on the phase of the driving laser field at the time it tunnels out of the potential. Plotting the total kinetic energy gain as a function of the driving electric field phase to find its maximum, this occurs at ~18° and ~$3.17U_p$, as seen in Fig. 3(b). If the electron evolves longer than this in the driving field, it can either return 180° out of phase or after an even number of optical cycles, which has an extremely low probability of occurrence.

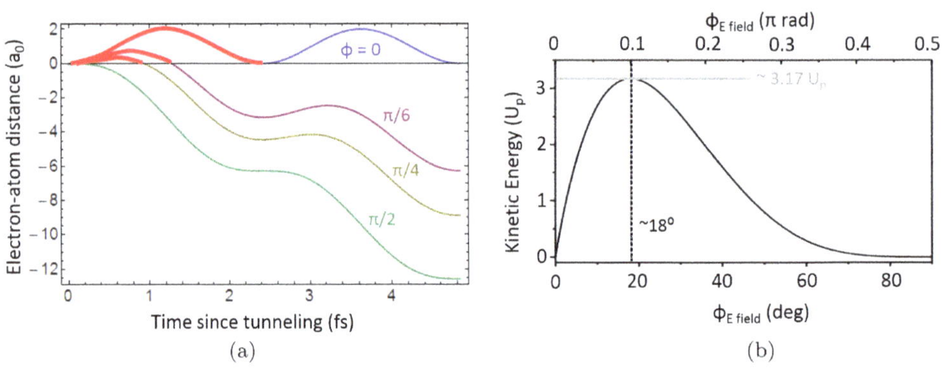

Fig. 3. (a) Plot of electron trajectory obtained from derived equations of motion. This trajectory is only "closed", returning to the ion (highlighted in red), when the electrons tunnels out between phases corresponding to 0–$\pi/2$ rad of the driving laser phase. (b) The kinetic energy gain versus the laser phase when the electron tunnels out of the potential. Adapted from Ref. [1].

3.5 Recombination

After gaining kinetic energy from accelerating in the laser field, the electron has a certain probability to recombine with the parent ion and emit a photon equal to the energy gained in the field. For an accurate model, other scattering processes need to be considered such as elastic scattering and collisional ionization. The total emission probability can be calculated via the dipole operator, taking into account the associated phase. The maximum emitted photon energy can be calculated via conservation of energy, where the amount of extra energy the electron has is dependent on the ionization potential of the atom and the maximum energy that it is able to gain from its evolution in the field:

$$h\nu_{\rm max} = I_p + 3.17U_p. \tag{9}$$

This serves as a maximum "cutoff" photon energy to our possible generated high harmonics [15]. As schematically shown in Fig. 4 [23], the resultant spectrum is seen to have three primary characteristics:

- An initial strong peak close to the fundamental wavelength
- A long plateau region with relatively equal intensity
- A sharp cutoff at high photon energies as $h\nu_{\rm max}$ is reached

In the time domain, HHG occurs every half cycle of the laser field in short attosecond bursts. The initial strong peaks in the emission are where the generated intensity can be modeled by perturbation theory. Relatively uniform intensity of the harmonics on the plateau is due to the efficiency in ionization being "nonperturbative" and relatively independent of generated harmonic order [13, 19]. In reality, these intermediate orders are not

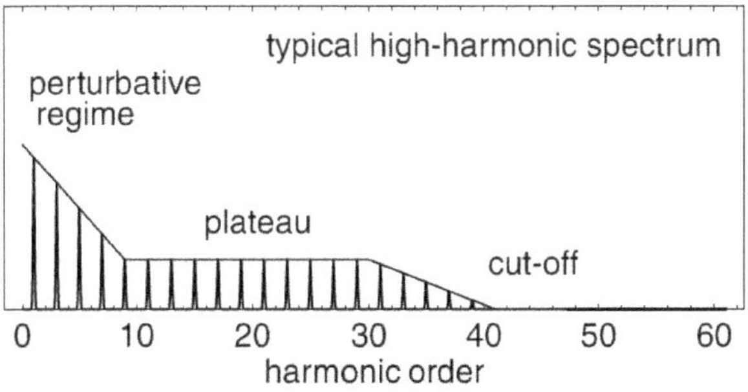

Fig. 4. Schematic showing the three key regions of HHG: low order, perturbative regime, plateau and cutoff (after Ref. [23]).

discretized. Since there are many electron trajectories that contribute to each harmonic order, each contribution to the order has a discrete frequency phase shift depending on when the emission of the HHG photon occurs. This leads to an interference effect that somewhat "smears" out the orders. The sharp cut off is then dictated by the limit in energy that an electron is able to gain from accelerating in the electric field after ionization.

The decrease in signal intensity at high energies results from a low number of electrons contributing to such high generated energies since these electrons would be ionized near the peak of the fundamental field. Also, due to this smaller number of electrons contributing to these orders, the interference effects are not as dramatic and the orders are more discretized. The overall conversion efficiency of the entire process is typically on the order of 10^{-5}–10^{-7} depending on the driving wavelength [6].

Krause *et al.* [15] have done studies on the harmonic cutoff and plateau properties shown in Fig. 5. High harmonic spectra are shown in He for 527 nm (solid line) and 1,053 nm (open circles) for 6×10^{14} W/cm^2 pulse intensity. Note that the cutoff for the 527 nm pulses occurs at a significantly lower energy than that for 1,053 nm, as a result of the $1/\lambda^2$ dependence of U_p, the ponderamotive potential.

As an example of harmonic plateaus generated with an intense 125 fs laser pulse some of the earliest work is shown in Fig. 6.

This work, carried out by Macklin *et al.* [24], used 15 mJ, 125 fs pulses of 806 nm light from an amplified Ti:sapphire laser. The peak intensity at

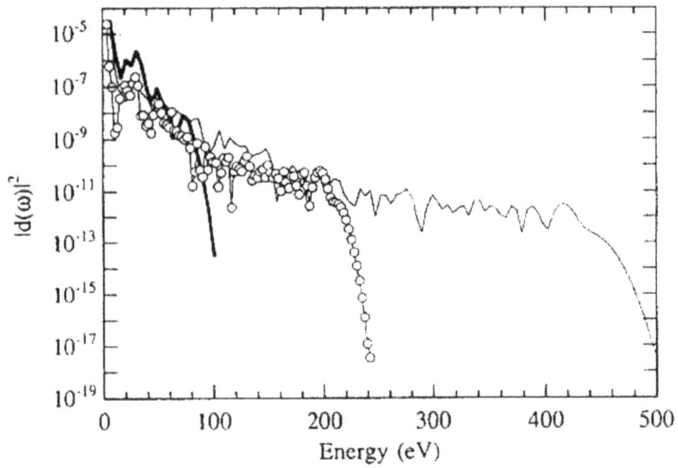

Fig. 5. High-harmonic spectra in He for 527 nm (solid line) and 1,053 nm (open circles) for 6×10^{14} W/cm^2 pulse intensity. Also shown (thin line) is the harmonic spectrum in He^+ at 527 nm but at 5×10^{15} W/cm^2 (after Ref. [15]).

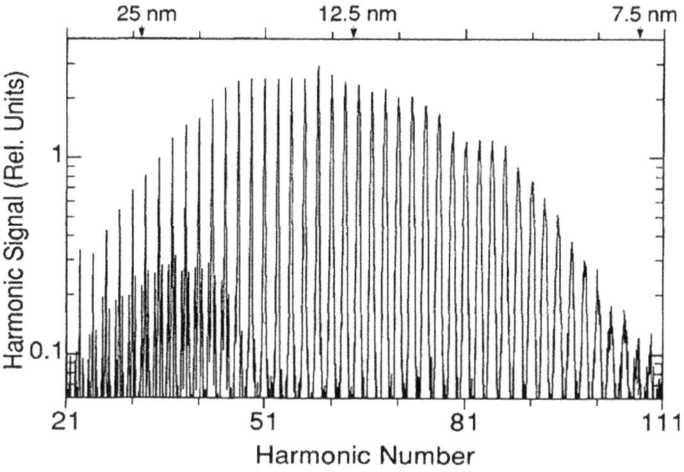

Fig. 6. Spectrum of harmonics generated in Ne for input pulse intensity of 1.2×10^{15} W/cm^2 (after Ref. [24]).

the focus was 1.3×10^{15} W/cm^2. The harmonics were generated by focusing the laser pulses into a tube with a small hole designed to confine the Ne gas to a pressure of 13 Torr.

3.6 HHG Characteristics

Without knowing anything of what the resultant spectrum looks like, some general observations can be made of what aspects of the process most drastically influence the overall harmonic emission output. From looking at the equations for the maximum cutoff energy and pondermotive potential, the highest generated wavelength is dependent on the laser peak intensity. Therefore, shorter laser pulses with higher peak intensity allow for shorter cutoff wavelengths since the electron "survives" for longer times in the field (to higher field strengths) before ionizing. This allows it to gain more kinetic energy from the pondermotive force. Another dependent factor is the ionization potential of the atom. Higher ionization potentials lead to higher cutoff frequencies. This is seen in Fig. 5 for He^+ whose ionization energy 54.4 eV. The influence of these two factors on the highest cutoff photon energy can be seen when looking at atoms like Neon and Helium, with $I_p = 21.6$ eV and 24.6 eV, respectively. With a 25 fs driving laser pulse width centered at 800 nm and 6×10^{15} W/cm^2 intensity, the highest harmonics possibly generated are the 163rd and 333rd (4.9 nm and 2.4 nm, respectively). By comparison, with a longer 100 fs pulse width assuming the same peak intensity, the cutoffs drop to the 119th and 237th harmonics for Neon and Helium, respectively [25].

Looking at the periodicity of the driving laser field, the HHG process also takes place every *half* cycle of the driving field, producing a series of attosecond bursts. Therefore, only *odd* harmonics are generated due to the odd symmetry of the generating process. To generate even harmonics, a medium without inversion symmetry would be required, making the harmonic emission always add constructively, no matter if the photon is generated from the field oscillating one way or the other. This does not exist for a gas. Introduction of a second pulse with a different fundamental wavelength could generate the even pulses via filling in the gaps of the harmonic spectrum.

3.7 Methods for Generating High Harmonics

There are a number of standard approaches for generating high harmonics. In this section we discuss two approaches, one utilizing a simple tube that confines the noble gas and a second that utilizes a piezoelectrically driven pulsed valve. In the next section, we discuss a third method for phase matching in hollow core glass tubes or capillaries that allow for significant enhancement of harmonic efficiency over a narrow range of harmonics that can be tuned through changes in geometry and gas pressure. An alternative method of using a "semi-infinite" gas cell has also gained popularity, with demonstrated conversion efficiencies in Ar around 10^{-5} [26, 27]. This relatively simple design consists of a confined gas-filled volume within a tube, defined between an entrance window and a thin membrane with an exit aperture roughly the size of the beam. We will focus our discussions to the tube/capillary confinement method and piezoelectric pulsed valve.

In the simpler approach described here, we discuss the two methods for confining the noble gas in a stainless steel tube and pulsed valve as shown in Fig. 7.

In both cases differential pumping, typically with a turbomolecular pump keeps the pressure within the vacuum beam line low. For the piezoelectric pulsed valve, pulse rates as high as 1 kHz are possible. In the standard geometry, shown in more detail in Chapter 4 when we discuss applications of the generated harmonics to photoelectron spectroscopy, the pulsed valve faces directly into the turbo pump for maximal containment of the noble gas. In essentially all cases, the co-propagating fundamental and harmonic light pulses are separated by a grating or multilayer mirror before use or analysis.

3.8 Phase Matching in a Capillary Waveguide

Up to this point, certain assumptions have been made about the characteristics of the high-harmonic light generated in the gas, the main assumption

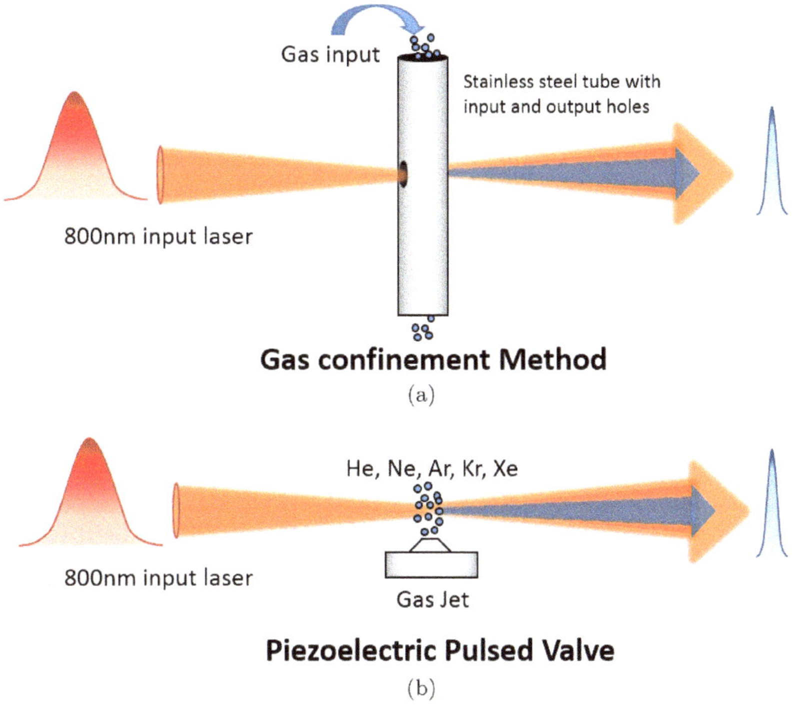

Fig. 7. (a) Schematic of a simple stainless steel tube to confine the noble gas with input (800 nm light pulse) and output (harmonics) holes. (b) Schematic of a pulsed valve that emits high pressure bursts of noble gas.

being that the light generated from an atom in one portion of the gas is exactly in phase with the light generated a certain distance later. If this is not the case, then the harmonic light does not add constructively and the output intensity is greatly reduced. Due to the significant difference in wavelengths between the incoming laser light and the generated EUV, there is an inherent phase mismatch between the two fields that must be accounted for. If we forcibly phase match our harmonic light with that of the input pulse wavelength, this can produce a 10^2–10^3 factor increase in output intensity compared to the non-phase matched case, allowing for greater experimental applications [28]. The characteristic length in which the phase of the harmonically generated light "slips" from the phase of the fundamental phase by π is called the *coherence length* and can vary from the mm scale for low (<150 eV) photon energy to micrometer scale for high (>200 eV) photon energy. A slip of the fundamental phase by π radians results in destructive interference between the input light (e.g. 800 nm) and harmonic light that results in a drop in the output harmonic intensity. Maximizing this coherence length allows for greater interaction length in which the harmonic light can constructively interfere, thereby producing an enhancement in the harmonic output.

Several factors contribute to the spatial "phase mismatch" that needs to be corrected for so that the output light is intense enough for practical applications. First, the gas in which the harmonic light is generated is a nonlinear medium. As the light propagates, it inherently picks up a phase lag such that

$$E_q \propto \int_0^L E_f^n d(z) e^{-i\Delta kz} dz, \tag{10}$$

propagating through a medium with length L, where E_q is the electric field of the qth harmonic, n is the order of the nonlinear process, E_f is the electric field of the fundamental, $d(z)$ is the nonlinear coefficient, Δk is the phase mismatch ($\Delta k = qk_f - k_q$) between the fundamental laser field wavevector, k_f, and the harmonic field k_q. Thus, for $\Delta k = 0$, E_q is maximized, corresponding to a maximum in output intensity. To see the extent of this effect on the final intensity, this expression can be integrated assuming E_f^n and $d(z)$ are independent of propagation direction. This gives

$$E_q \propto L^2 \text{sinc}^2 \left(\frac{\Delta kL}{2} \right). \tag{11}$$

With no phase mismatch, the output harmonic signal grows as L^2, serving as a strong motivation to reduce the mismatch to zero. If Δk is non-zero then it oscillates over the propagation distance with the fields slipping in and out of constructive interference.

Given how strongly the phase mismatch effects the output intensity, several phase matching approaches have been attempted to minimize Δk. Expressing this mismatch more precisely, three district components contribute to the overall phase difference:

$$\Delta k_{\text{total}} = \Delta k_{\text{disp}} + \Delta k_{\text{plasma}} + \Delta k_{\text{geom}}, \tag{12}$$

where Δk_{disp} is the dispersion due to propagation in the neutral gas medium, Δk_{plasma} is the dispersion in the plasma created from the unrecombined free carriers in the generating medium, and Δk_{geom} is the geometrical dispersion when confined in, for example, a waveguide. One of the most successful approaches at minimizing Δk_{total} has been by using a gas-filled capillary waveguide.

In this configuration, the total Δk expression can be written in which the inherent waveguide dispersion is included as

$$\Delta k_{\text{total}} = [n(\omega_f) - n(m\omega_f)]\frac{\omega_f}{c} + \frac{\omega_p^2(1-m^2)}{2mc\omega_f} + \frac{u_n l^2 c(1-m^2)}{2ma^2\omega_f}, \tag{13}$$

where ω_f is the fundamental laser frequency, $n(\omega)$ is the frequency dependent index of refraction of the medium, m is the mth harmonic order, ω_p is the plasma frequency, a is the inner radius of the capillary and $(u_n)l^2$ is the lth zero of the Bessel function $J_{n-l}(u_{\text{nl}}) = 0$ [19]. In general, the positive dispersion of the index of refraction term k_{disp} can be controlled to cancel the negative dispersions of the plasma (Δk_{plasma}) and waveguide (Δk_{geom}). A final phase-matched harmonic signal can then most easily be achieved by tuning the gas pressure within the capillary to adjust the density of the neutral medium.

Capillary-based phase matching [28] is shown in Fig. 8 where femtosecond 800-nm light pulses are focused at the input of a hollow capillary. The capillary is sectioned for introduction of the noble gas employed in the harmonic conversion and for pressure tuning to achieve phase matching.

Figure 9 shows the results of pressure tuning to achieve phase matching [29] from the approach described above. Figure 9(a) shows the intensity dependence upon Ar pressure within the capillary for the 19th, 21st, and 23rd harmonics of the 800-nm input light pulse. In Fig. 9(b), the phase matched enhancement of the 27th and 29th harmonics are shown at an Ar pressure of 30 Torr. The peaking of those harmonics shows the effect on intensity of the

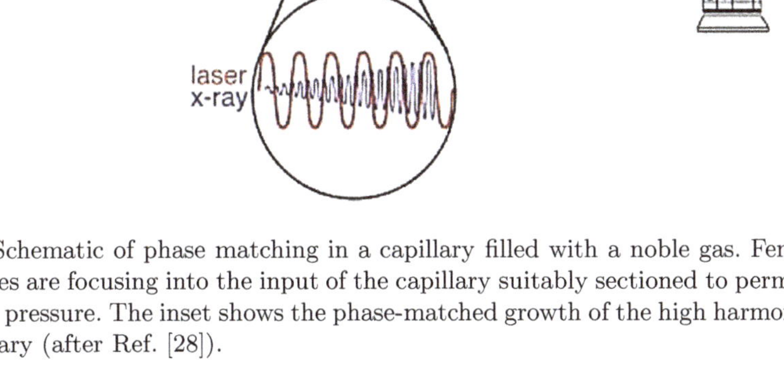

Fig. 8. Schematic of phase matching in a capillary filled with a noble gas. Femtosecond laser pulses are focusing into the input of the capillary suitably sectioned to permit control of the gas pressure. The inset shows the phase-matched growth of the high harmonic within the capillary (after Ref. [28]).

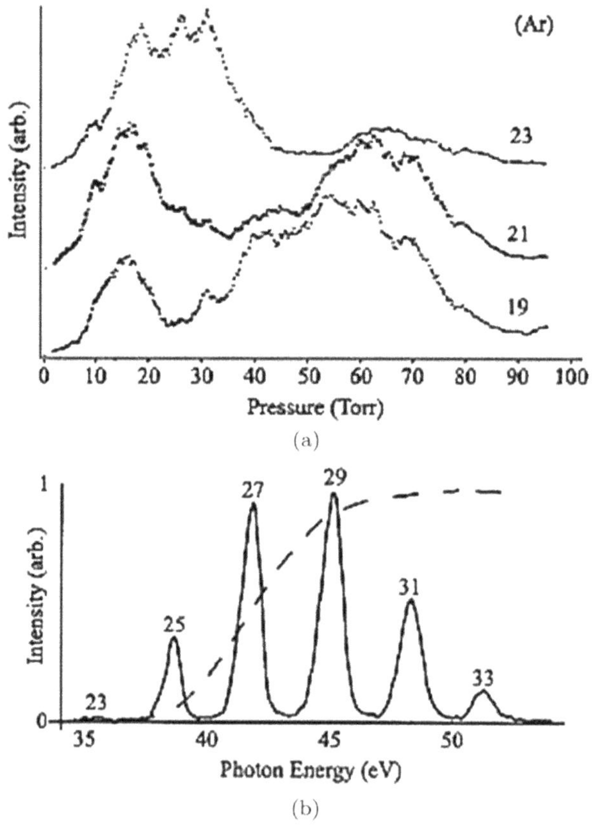

Fig. 9. (a) Dependence of harmonic intensity on Ar pressure for the 19th, 21st, and 23rd harmonics of 800 nm light showing the peaks in emission where phase matching of the input and harmonic light occur. (b) Intensity of a narrow band of phase matched harmonics at an Ar pressure of 30 Torr (after Ref. [29]).

phase matched harmonics (constructive interference at those wavelengths) compared with adjacent harmonics <25 and >31, where those harmonics are suppressed indicating destructive interference.

Clearly there are many different approaches to generating high harmonics, ranging from confined gas geometries to piezoelectrically driven pulsed valves that increase the instantaneous gas density while minimizing gas pressure outside the focal region. These relatively straightforward approaches provide for the generation of substantial fluxes of tunable high harmonics that can be utilized for spectroscopic and metrological applications.

This chapter described the fundamental physics of HHG as well as the characteristics of the generated short wavelength light. We described the methods by which harmonics are generated including non-phase matched and phase matched approaches to enhance particular harmonic emission. In Chapter 4, we next describe particularly powerful applications of these harmonics in photoelectron spectroscopies that reveal the electronic properties and dynamics of materials.

References

[1] T. Pfeifer, C. Spielmann, and G. Gerber, "Femtosecond X-ray science," *Reports Prog. Phys.*, vol. 69, no. 2, pp. 443–505, 2006.

[2] C. D. Macchietto, B. R. Benware, and J. J. Rocca, "Generation of millijoule-level soft-X-ray laser pulses at a 4-Hz repetition rate in a highly saturated tabletop capillary discharge amplifier," *Opt. Lett.*, vol. 24, no. 16, pp. 1115–1117, 1999.

[3] J. J. Rocca, V. Shlyaptsev, F. G. Tomasel, O. D. Cortazar, D. Hartshorn, and J. L. A. Chilla, "Demonstration of a discharge pumped table-top soft-x-ray laser," *Phys. Rev. Lett.*, vol. 73, no. 16, p. 2192, 1994.

[4] A. McPherson, G. Gibson, H. Jara, U. Johann, T. S. Luk, I. A. McIntyre, K. Boyer, and C. K. Rhodes, "Studies of multiphoton production of vacuum-ultraviolet radiation in the rare gases," *JOSA B*, vol. 4, no. 4, pp. 595–601, 1987.

[5] A. Rettenberger and R. Haight, "Multivalley electron population dynamics on the Ge (111): As surface," *Phys. Rev. Lett.*, vol. 76, no. 11, p. 1912, 1996.

[6] T. Popmintchev, M.-C. Chen, D. Popmintchev, P. Arpin, S. Brown, S. Ališauskas, G. Andriukaitis, T. Balčiunas, O. D. Mücke, and A. Pugzlys, "Bright coherent ultrahigh harmonics in the keV X-ray regime from mid-infrared femtosecond lasers," *Science (80-.).*, vol. 336, no. 6086, pp. 1287–1291, 2012.

[7] T. Popmintchev, M.-C. Chen, P. Arpin, M. M. Murnane, and H. C. Kapteyn, "The attosecond nonlinear optics of bright coherent X-ray generation," *Nat. Photonics*, vol. 4, no. 12, pp. 822–832, 2010.

[8] S. Mathias, S. Eich, J. Urbancic, S. Michael, A. V. Carr, S. Emmerich, A. Stange, T. Popmintchev, T. Rohwer, and M. Wiesenmayer, "Self-amplified photo-induced gap quenching in a correlated electron material," *Nat. Commun.*, vol. 7, 2016.

[9] R. Haight, "Photoemission with laser-generated harmonics tunable to 80 eV," *Appl. Opt.*, vol. 35, no. 33, pp. 6445–6448, 1996.

[10] R. Haight and D. R. Peale, "Tunable photoemission with harmonics of subpicosecond lasers," *Rev. Sci. Instrum.*, vol. 65, no. 6, pp. 1853–1857, 1994.

[11] R. Haight and D. R. Peale, "Antibonding state on the Ge (111): As surface: Spectroscopy and dynamics," *Phys. Rev. Lett.*, vol. 70, no. 25, p. 3979, 1993.

[12] N. Sarukura, K. Hata, T. Adachi, R. Nodomi, M. Watanabe, and S. Watanabe, "Coherent soft-X-ray generation by the harmonics of an ultrahigh-power KrF laser," *Phys. Rev. A*, vol. 43, no. 3, p. 1669, 1991.

[13] M. Lewenstein, P. Balcou, M. Y. Ivanov, A. L'huillier, and P. B. Corkum, "Theory of high-harmonic generation by low-frequency laser fields," *Phys. Rev. A*, vol. 49, no. 3, p. 2117, 1994.

[14] W. Becker, A. Lohr, M. Kleber, and M. Lewenstein, "A unified theory of high-harmonic generation: Application to polarization properties of the harmonics," *Phys. Rev. A*, vol. 56, no. 1, p. 645, 1997.

[15] J. L. Krause, K. J. Schafer, and K. C. Kulander, "High-order harmonic generation from atoms and ions in the high intensity regime," *Phys. Rev. Lett.*, vol. 68, no. 24, p. 3535, 1992.

[16] T. D. Donnelly, T. Ditmire, K. Neuman, Md. Perry, and R. W. Falcone, "High-order harmonic generation in atom clusters," *Phys. Rev. Lett.*, vol. 76, no. 14, p. 2472, 1996.
[17] J. Itatani, J. Levesque, D. Zeidler, H. Niikura, H. Pépin, J.-C. Kieffer, P. B. Corkum, and D. M. Villeneuve, "Tomographic imaging of molecular orbitals," *Nature*, vol. 432, no. 7019, pp. 867–871, 2004.
[18] D. Von der Linde, T. Engers, G. Jenke, P. Agostini, G. Grillon, E. Nibbering, A. Mysyrowicz, and A. Antonetti, "Generation of high-order harmonics from solid surfaces by intense femtosecond laser pulses," *Phys. Rev. A*, vol. 52, no. 1, p. R25, 1995.
[19] F. Pfeiffer, T. Weitkamp, O. Bunk, and C. David, "Phase retrieval and differential phase-contrast imaging with low-brilliance X-ray sources," *Nat. Phys.*, vol. 2, no. 4, pp. 258–261, 2006.
[20] M. V. Ammosov, N. B. Delone, and V. P. Krainov, "Tunnel ionization of complex atoms and of atomic ions in an alternating electromagnetic field," *Sov. Phys. JETP*, vol. 64, no. December 1986, pp. 1191–1194, 1986.
[21] O. Kfir, P. Grychtol, E. Turgut, R. Knut, D. Zusin, D. Popmintchev, T. Popmintchev, H. Nembach, J. M. Shaw, and A. Fleischer, "Generation of bright phase-matched circularly-polarized extreme ultraviolet high harmonics," *Nat. Photonics*, vol. 9, no. 2, pp. 99–105, 2015.
[22] D. D. Hickstein, F. J. Dollar, P. Grychtol, J. L. Ellis, R. Knut, C. Hernández-García, D. Zusin, C. Gentry, J. M. Shaw, and T. Fan, "Non-collinear generation of angularly isolated circularly polarized high harmonics," *Nat. Photonics*, 2015.
[23] C. Winterfeldt, C. Spielmann, and G. Gerber, "Colloquium: Optimal control of high-harmonic generation," *Rev. Mod. Phys.*, vol. 80, no. 1, p. 117, 2008.
[24] J. _J. Macklin, J. D. Kmetec, and C. L. Gordon III, "High-order harmonic generation using intense femtosecond pulses," *Phys. Rev. Lett.*, vol. 70, no. 6, p. 766, 1993.
[25] Z. Chang, A. Rundquist, H. Wang, M. M. Murnane, and H. C. Kapteyn, "Generation of coherent soft X-rays at 2.7 nm using high harmonics," *Phys. Rev. Lett.*, vol. 79, no. 16, p. 2967, 1997.
[26] Y. Oishi, M. Kaku, A. Suda, F. Kannari, and K. Midorikawa, "Generation of extreme ultraviolet continuum radiation driven by a sub-10-fs two-color field," *Opt. Express*, vol. 14, no. 7230, 2006.
[27] E. Takahashi, Y. Nabekawa, T. Otsuka, M. Obara, and K. Midorikawa, "Generation of highly coherent submicrojoule soft X-rays by high-order harmonics," *Phys. Rev. A*, vol. 66, no. 21802, 2002.
[28] A. Rundquist, C. G. Durfee, Z. Chang, C. Herne, S. Backus, M. M. Murnane, and H. C. Kapteyn, "Phase-matched generation of coherent soft X-rays," *Science (80-.).*, vol. 280, no. 5368, pp. 1412–1415, 1998.
[29] C. G. Durfee III, A. R. Rundquist, S. Backus, C. Herne, M. M. Murnane, and H. C. Kapteyn, "Phase matching of high-order harmonics in hollow waveguides," *Phys. Rev. Lett.*, vol. 83, no. 11, p. 2187, 1999.

Chapter 4

Femtosecond Photoelectron Spectroscopy: Fundamentals and Electron Dynamics

4.1 Introduction

Photoelectron spectroscopy (photoemission) has a long history originating with Einstein's initial explanation of the photoelectric effect in solids [1]. But we would not start that far back. The interaction of energetic light with materials, whether they be solids, gases, or liquids is of fundamental importance in physics. Photoemission allows us to peer into the electronic interior of these materials in a unique way. When combined with structural characterization that includes, for example, electron microscopies, atomic force microscopies, ion scattering, and more, a full characterization of a material and its properties is within our reach.

In particular, photoemission provides an in-depth view of the electronic structure and properties of materials. Energy states occupied by electrons within the valence bands of materials, known as its electronic band structure, can be elucidated with angle resolved photoemission. Chemical states of atoms at and near the surface can also be interrogated by investigating the deeper bound core levels that shift in response to charge transfer resulting from chemical reactions. Collective electronic properties that include band edge and Fermi level location, the alignment of states at the interface between two dissimilar materials, the work function of materials under various conditions, magnetic properties and more can all be interrogated with photoemission spectroscopy.

Photoelectron spectroscopy can be separated, perhaps a bit arbitrarily, into two energetic regimes. The first regime extends from $\sim$5 eV, corresponding to the minimum energy required to liberate an electron from a solid (work function) to $\sim$1 keV. Photoelectron spectroscopy is typically referred to as ultraviolet photoemission spectroscopy (UPS). The lowest photon energies correspond to the typical work functions of most materials of $\sim$4–5 eV although the alkali metals and alkali-dressed solid surfaces can have work functions as low as ~3 eV. 5 eV photons are not absorbed by atoms comprising air but beyond ~6–7 eV, these photons are increasingly absorbed in air and are referred to as the vacuum ultraviolet. Photons with energies beyond 1 keV are used for X-ray photoelectron spectroscopy, also known as XPS. At these higher energies, light penetrates solids more effectively, photoexcited valence electrons have longer inelastic mean-free paths (as will be discussed later in this chapter) and atomic core levels for most, if not all, elements are energetically accessible.

In XPS, energetic photons, typically in the range of several keV can photoemit deeply bound core level electrons from within atoms. While core electrons are not involved in chemical bonding directly, their binding energies can shift as a result of valence electron transfer in a chemical reaction between elements. This will be discussed in greater detail in Section 4.4.

In UPS, photons below 1 keV but typically in the range of 20–300 eV are used to photoemit electrons from shallow core levels and valence states of materials. In crystalline materials, detailed analysis of the electron emission direction can be used to determine its wavevector within the solid [2]. As a result the electronic band structure of the crystal can be derived. This process has been extensively studied and a number of excellent review articles can be found in the literature. We describe the core concepts below [3].

The fundamental understanding of electron excitation in a solid by an energetic photon can be traced to Einstein [1], who was awarded Nobel Prize in 1921 for delineating this simple but exceedingly important process. Photoelectron emission experiments revealed the quantized nature of individual photons and the energy they carry by demonstrating that while short wavelength, UV light liberates electrons from gases and solids, infrared light does not. This behavior is independent of intensity. Of course, extraordinarily intense femtosecond pulses of infrared light from lasers can emit electrons through the nonlinear process of multiphoton absorption but it would take nearly 40 more years before lasers were even invented!

In solids, the photoemission process can be resolved into a sequence of events which begin with the absorption of an energetic photon that promotes an electron to a state above the vacuum level of the solid (Fig. 1). The vacuum level is that energy to which a liberated electron exists outside the solid with zero kinetic energy. Therefore, the kinetic energy of the electron, measured by an electron analyzer is given by,

$$E_{\text{kin}} = h\nu - (E_{\text{vac}} - E_i) + E_{\text{cp}}, \tag{1}$$

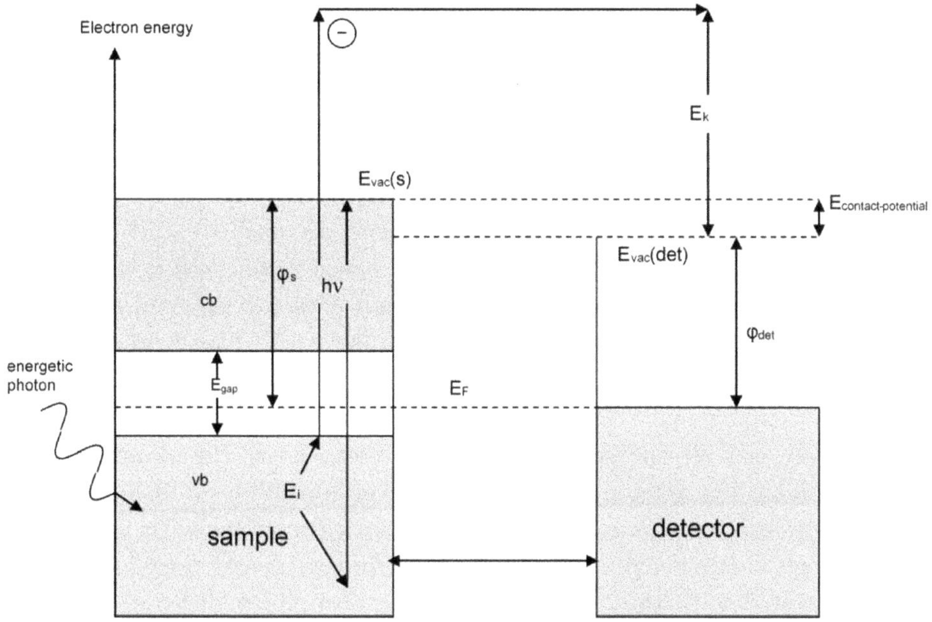

Fig. 1. Schematic showing fundamental energetic considerations in photoelectron spectroscopy. Connection of the sample to the detector equilibrates Fermi levels, but will introduce a contact potential.

where E_i is the initial state energy the electron occupies within the solid, typically in the valence band of the solid, but also within the conduction band if the electron exists there due to photoexcitation or doping. If the electron is at the Fermi level, as in the case of a metal, $(E_{\text{vac}} - E_i)$ would correspond to the work function, ϕ, of the solid. E_i is the binding energy with respect to the Fermi level and E_{cp} is the contact potential created when the analyzer is electrically connected to the sample.

In photoelectron spectroscopy, the Fermi level is the fundamental reference level. A useful paper on these considerations by Cahen and Kahn [4] provides a thorough discussion of photoemission energetics, but we capture the salient points here. When two unrelated solids are brought into intimate electrical contact (implying that electrons can flow from one solid to the other) electrons will flow from the solid with the smaller work function to the solid with the larger work function until the Fermi levels of the two solids equilibrate or are the same.

A simple analogy is the connection of two buckets of water filled to different levels. Separately, each bucket's level (Fermi level) is different. If we punch a hole in each bucket and connect a tube between them, water will flow until the levels are the same in each bucket. Hence, their "Fermi level" has equilibrated. In the case where water is replaced with electrons, while the Fermi levels are now equal, electronic charge is transferred.

The result of this charge transfer is that an electrostatic potential is then set up between the two materials, called a contact potential. Interestingly, this occurs between two materials in intimate contact or between a solid under study and the electron energy analyzer used to study it within photoemission. In the case of the material couple, the field is internal to the coupled solid while for a solid and electron detector the field exists as a voltage difference between the solid and the detector. This field can accelerate or slow the photoemitted electrons and must be taken into account during the experiment. Nonetheless, the Fermi level is then the key reference level and all energetic changes experienced by the sample during the experiment are referenced to it. This level is easily established to high precision by evaporating a metal film in electrical contact with the sample and then generating a photoemission spectrum to reveal the sharp Fermi edge associated with the cutoff of electronic occupancy dictated by Fermi statistics [5].

Once the Fermi level of the system is established, it is the level to which all else is referred. This is particularly important since, as we will discuss in later chapters, femtosecond pump-probe photoelectron spectroscopy (which we will call fs-UPS) can also be used to study dynamic systems that may include changes in band bending in the materials under study [6]. Relating spectroscopic information to the Fermi level of the system is then key to extracting the electronic structure of systems under study such as band bending, band offsets in heterostructures and the locations of the valence band maxima and conduction band minima.

4.2 Electron Escape Depth

Another set of important considerations, particularly relevant to photoemission from a nanowire (NW) or other nanoscale structures involves the photoemission photon energy and the inelastic mean-free path of the emitted electrons [7]. The absorption of light, of intensity Io, in a material is given by

$$I = I_o e^{-\alpha z}, \tag{2}$$

where α is the optical absorption coefficient and z is the distance into the solid. A useful tabulation of absorption coefficients for photon energies up to 6 eV, for a range of elemental and compound semiconductors can be found in a paper by Aspnes and Studna [8]. As we will discuss later, the absorption of

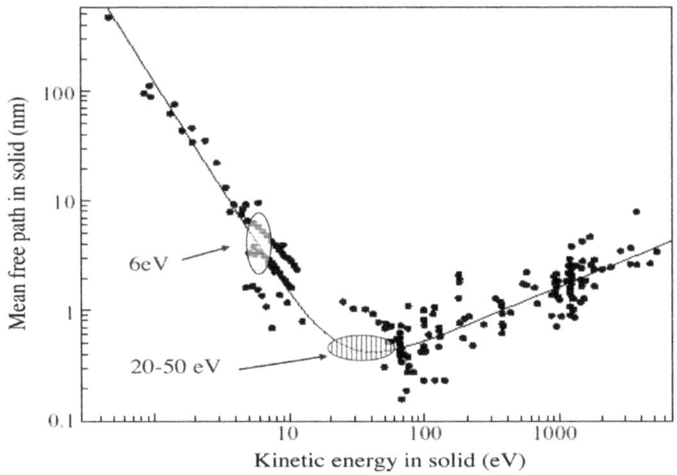

Fig. 2. Electron mean-free path as a function of kinetic energy of the electron in the solid. For 6 eV electrons the mean-free path is approximately 6–10 nm [7].

6 eV photons in Si results in an intensity drop to its 1/e value in ~6–10 nm from the surface. The inelastic mean-free path for electrons with an energy of 6 eV above the Fermi level is ~6–10 nm (Fig. 2.)

4.3 Angle Integrated, Angle Resolved, and X-ray Photoelectron Spectroscopy

In photoelectron spectroscopy, there is an array of different approaches including angle resolved and angle integrated spectroscopies and XPS. Each is applied to study specific properties of the material. For example, angle resolved UPS (ARUPS) is used to study the energy versus wavevector electronic band structure of single crystal semiconductors and metals [3]. In ARUPS, the photon energies used range typically from 10 eV to ~300 eV and often the polarization state of the photons, used to elicit fundamental symmetries, is well defined. But when, for example a polycrystalline material is studied the specific energy–wave vector information is lost due to averaging over the random orientations of the crystal grains within the material. This is also true for amorphous materials where long-range order is absent. Under those situations an angle integrated detector is often the most useful and the resultant photoelectron spectra can be interpreted in terms of a total density of states as a function of binding energy. These spectra can then be compared, for example, with theoretical calculations such as density functional theory (DFT) [9] that can generate information on state density as a function of binding energy.

4.4 X-ray Photoelectron Spectroscopy

Moving into the realm of more energetic X-ray photons, XPS is carried out with photons of substantially higher energies where the deeply bound, elementally specific atomic core electrons are accessible. The photoionization cross-sections [10] are accurately known and when coupled with the electron escape depth of the photoemitted electron, the elemental concentration at and near the surface of a material under study can be determined.

Swedish scientist, Kai Siegbahn [11] discovered that atomic core levels shift rigidly in energy when the atom is involved in chemical bonding and that these shifts can be observed in X-ray photoelectron spectra and can be directly related to the oxidation state of the atom involved in binding. Electronic charge transfer from cation to anion results in a shift of the cation atomic core level to higher binding energy and lower binding energy for the anion. This affect was first demonstrated and explained by Siegbahn, for which he won the Nobel Prize in Physics (1981) and is now routinely used to identify the oxidation states and chemical properties of material surfaces.

This shift is a consequence of fundamental electrostatics. Atomic core level binding energies are determined by the electrostatic interaction between the core electrons and the nucleus whose charge is fixed, with additional contributions (screening) from the valence electrons. Removal or addition of valence charge due to a chemical reaction will shift the core level binding energy. If we make the somewhat crude approximation that the valence electrons can be thought of as a hollow charged conducting sphere at a radius R from the nucleus, then the contribution of the valence electrons to the *electric* field within the sphere is zero. The *potential* inside the sphere is a constant given by $V_{\text{inside}} \sim Q/4\pi\varepsilon_o R$, where Q is the valence charge. If a chemical bond is formed in which the bond has ionic character, i.e. if Q changes due to a transfer of an electron from one atom to another in a chemical reaction, V_{inside} changes linearly with the loss or gain of valence charge. Of course, the nuclear charge does not change and so a loss of a valence electron in a reaction (oxidation) will increase the binding energy of a core electron, while the addition of valence charge (reduction) will decrease the binding energy.

A typical example [12] and one of great industrial interest is the oxidation of Si. The $2p$ electrons of Si are found at a binding energy of 100 eV below its valence band maximum (vbm). Typically, binding energy is defined relative to the Fermi level of a solid; in semiconductors, the Fermi level can vary over the band gap depending on the type and degree of doping so it is defined relative to the top of the valence band in this case. The formation of SiO_2 results in transfer of electronic charge from the relatively electropositive

Si to the more electronegative oxygen atoms. The $2p$ core electrons in Si shift to higher binding energy in response to this transfer of charge from Si to O. As mentioned above, Kai Siegbahn [11] was the first to recognize and describe this effect and the XPS or electron spectroscopy for chemical analysis (ESCA) technique was born. XPS is now a well-established analytical technique and entire systems can be purchased commercially.

As an example, consider a fundamentally important industrial material: Si. SiO_2 is nature's gift to the microelectronics industry as the key gate insulator in metal–oxide–semiconductor (MOS) transistors for nearly 40 years of development. The most stable Si oxide is SiO_2, but other suboxides can form. Evidence of this is found in a paper that studied the detailed oxides formed at the Si–SiO_2 interface [12].

As can be seen from Fig. 3, the Si $2p$ core level for unoxidized Si is found at an initial state energy of −99.5 eV. Fully oxidized Si in the form of SiO_2 is

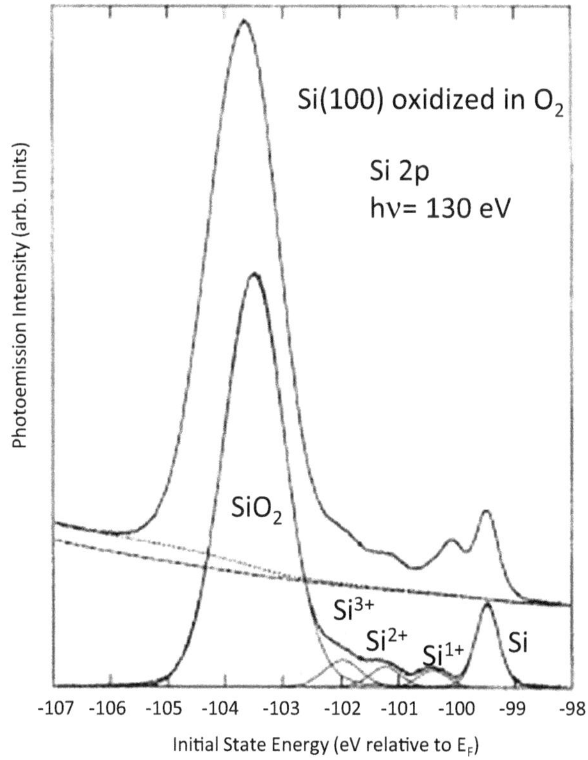

Fig. 3. XPS studies of the Si–SiO_2 interface showing the multiple suboxides existing on this surface. After background subtraction and careful peak fitting, the individual oxidation states of Si are resolved ranging from Si° to Si^{4+} [12]

found at an energy of $-103.7\,\text{eV}$. The $\sim$4 eV shift of the Si $2p$ core level for Si atoms bonded to 2 oxygens is due to the transfer of 4 Si sp^3 valence electrons to the highly electronegative O atoms satisfying the octet rule in chemistry. As a result of this charge transfer, an increase in the potential well shifts the inner shell $2p$ core electrons to *higher* binding energy (greater initial state energy in Fig 3). Conversely, since the O atoms pick up electrons from the Si atom the O 1s core level shifts to *lower* binding energy. Of additional interest is the emission intensity between the Si^0 peak at $-99.5\,\text{eV}$ and the fully oxidized Si^{4+} peak in SiO_2 found at $-103.7\,\text{eV}$. The structure of this emission is real and can be resolved into individual peaks corresponding to increasing oxidation states from Si to $Si^{1+,2+,3+}$, and finally Si^{4+}. Each peak corresponds to sub-oxides where Si has transferred either 0, 1, 2, 3, or 4 electrons to the oxygens. We note that for each increasing oxidation state of Si, the peak shifts by $\sim$0.5 eV. XPS carried out in this fashion can tell us the detailed chemistry occuring at this interface and more broadly it can reveal detailed chemistry whenever elements react and transfer electrons between atoms. Additionally, the intensity of each peak, when deconvolved, is directly related to the number of atoms involved in the reaction, so XPS is quantitative both in the energy of the core level shifts as well as atom numbers.

As mentioned, a broader term, ESCA has been coined for this technique and stand-alone, commercially available tools using single characteristic line sources such as Mg:Kα (1253.6 eV) and Al:Kα (1486.6 eV) are available. These sources have a natural linewidth of $\sim$0.4 eV limiting the overall resolution of the technique; as a result, monochromatized sources or synchrotrons are often used to achieve higher energy resolution. XPS/ESCA is an extraordinarily useful technique as it is not only used to extract the concentration of elemental species that comprise the surface of a material, but also reveals the chemical states of the elements as well.

4.5 Angle-Resolved UPS

There are a number of excellent review articles on angle resolved photoelectron spectroscopy [3, 6] (and references therein) but a brief description is given here. When a photon whose energy exceeds the work function (metal) or ionization energy (semiconductor or insulator) is incident upon a single crystal surface, the photoemitted electron carries specific information on the relationship between its energy and wavevector, or electronic band structure within the crystal. We can naturally break up angle resolved photoelectron spectroscopy into two general areas; the three-dimensional bulk and the two-dimensional surface. Photoelectron emission from bulk electronic states can be essentially resolved into a sequence of steps [13] that begin with the absorption of a photon sufficiently energetic to promote the electron to a state above the vacuum level (Fig. 1). At this point the kinetic energy of the electron within the crystal is:

$$E_{\text{kin}} = h\nu - (E_{\text{vac}} - E_i) \tag{3}$$

where E_{i} is the initial-state energy of the electron within the band structure. If E_i was at the Fermi level, E_F, then $E_{\text{vac}} - E_i$ is the work function ϕ of the material. If all energies are referenced to the Fermi level, which is a constant of the static system, then any state below E_F is characterized by a binding energy. Since ultrafast lasers make it possible to photoexcite electrons above the Fermi level into unoccupied states of the system, whether they be conduction band states, interface states or defect levels, we must also consider electrons with a negative binding energy.

Upon absorption of the ionizing photon, the bulk electron must transit to and pass through the surface where it then couples to a vacuum plane-wave state. The inelastic mean-free path of the electron, as described above, limits the depth from which an electron can originate and carry information about its original state to the electron detector. Inelastic scattering can carry information about phonons in the system or other energy loss (or gain) processes such as plasmons that are collective excitations of the electron gas within the material. There is a range of such inelastic collisions that the excited electron can undergo, and such interactions have been studied extensively [14]. Typically though, electrons may scatter inelastically many times as they exit the material and ultimately contribute to the broad background observed in photoelectron spectroscopy.

The transition rate between two eigenstates in the presence of a photon field is given by Fermi's Golden Rule

$$W = (2\pi/h)|\langle\psi_{\text{f}}|H_\nu|\psi_{\text{i}}\rangle|^2\, \rho(E_f)\, \delta(E_f - E_i), \tag{4}$$

where $H_\nu = -(\text{e}/\text{m})\mathbf{A}\cdot\mathbf{p}$. In this equation, $\mathbf{A}\cdot\mathbf{p}$ is the dipole approximation Hamiltonian for an electron in an electromagnetic field, where $\mathbf{A}$ is the Lorentz gauge vector potential. Importantly, $\mathbf{A}$ carries the polarization of the energetic photon and $\mathbf{p}$ is the momentum operator. $\rho(E_f)$ is the density of final sates to which the electron is excited from its initial state by the energetic photon. The delta function insures energy conservation.

There is a large amount of information stored in this seemingly simple equation and suggests experimental parameters that can be varied to increase knowledge about the band structure in the material. For example, since **A** carries the photon polarization, states which are excited by a photon with a particular polarization reveal the symmetry of the initial state. Selection rules [15], in analogy with light absorption in gases are polarization dependent.

As an example, consider electrons occupying dangling bond states at a semiconductor surface. If the dangling bonds extend along the surface normal then the dot product in the Hamiltonian requires that the electric field vector of the energetic photon possesses significant projection along the surface normal to ensure emission of electrons from the dangling bond states. For light parallel to the surface, the dot product is zero and little or no emission is observed. Figure 4 displays the experimental geometry of the incident light and electron detector.

When using ultrafast lasers to generate high-harmonics [16, 17] (vacuum ultraviolet and soft X-rays), the linear polarization [18] direction relative to the surface normal of the sample can be selected *prior* to upconverting the light for UPS. Relatively simple optical components such as half-wave plates designed for the visible and infrared regions of the spectrum where femtosecond, Ti:sapphire-based lasers operate, are commercially available.

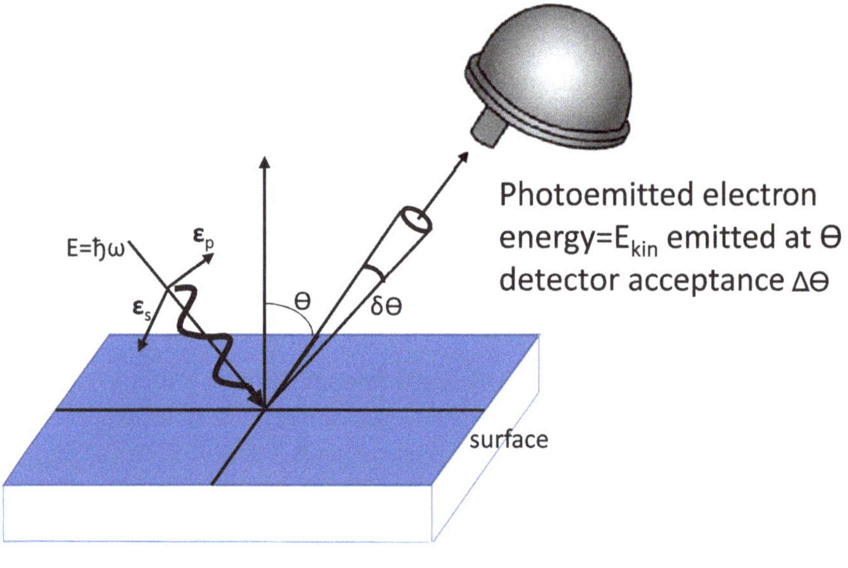

Fig. 4. Schematic of ARUPS. An incoming photon with a given electric field polarization $\varepsilon_\mathbf{p}$ (mixed perpendicular and parallel to the surface) or $\varepsilon_\mathbf{s}$ (purely parallel to the surface) impinges on the surface of a solid. A photoemitted electron is emitted at angle Θ with a detector acceptance angle $\delta\Theta$.

The availability of these optical components facilitates straightforward control over the polarization state of the light. As shown in Fig. 4, the photon electric field polarization can be located purely parallel to the surface ε_s, or as a mixture of parallel and perpendicular ε_p. Control of the polarization can be exploited to excite or inhibit photoemission of specific states, revealing detailed symmetries.

In the bulk of the material, similar symmetries exist and group theory character tables [15], representative of the symmetry of the specific crystal can be used to predict which transitions are allowed and which are forbidden based on the polarization of the incident light.

As can be seen from Fig. 5, tunability of the photoemitting light is important since coupling initial and final states within the Brillouin zone requires specific photon energies [19]. All transitions are vertical within the reduced Brillouin zone as vacuum UV photons carry nearly zero momentum. Further constraints result from the conservation of momentum of the photoemitted electron as it couples to plane waves outside the sample. Optical transitions occur from an initial state within the solid to a free electron final state as

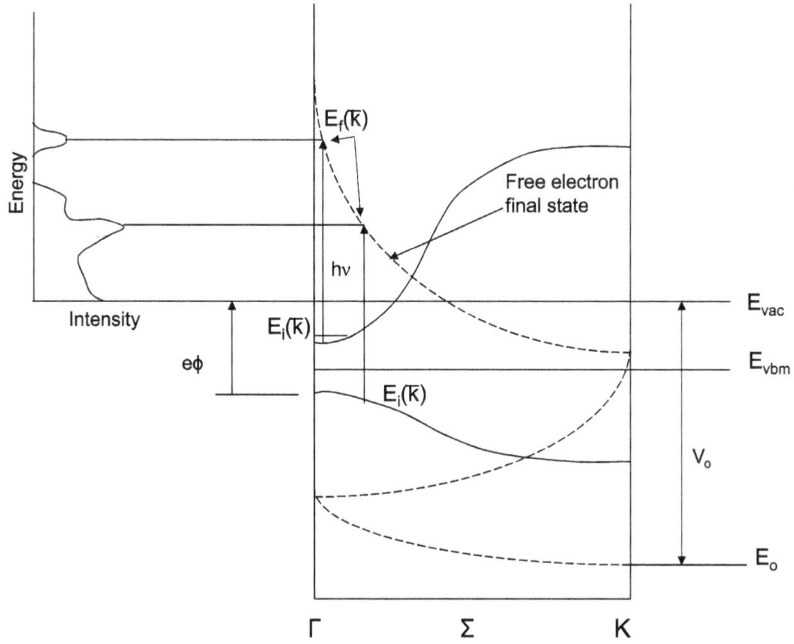

Fig. 5. Energy versus wavevector in the reduced Brillouin zone schematic of a crystalline semiconductor shown along a particular line of symmetry. The zone center is at Γ. The dashed lines are the free-electron band whose kinetic energy zero corresponds to E_o. Vertical transitions couple the initial state within the solid to a final state free electron band as shown.

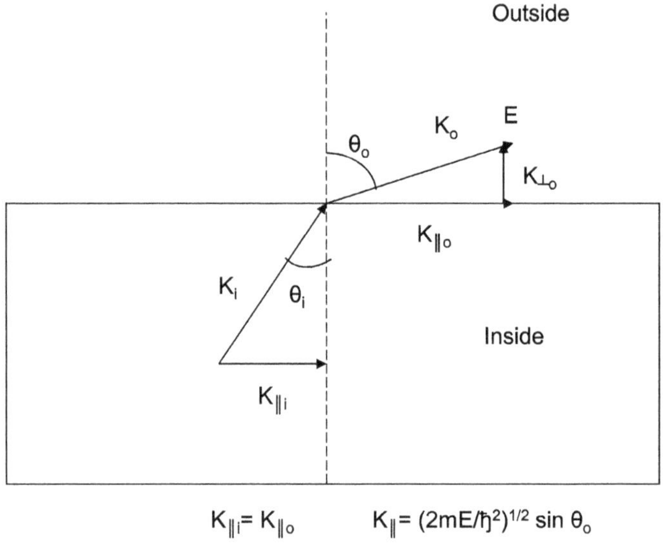

Fig. 6. Schematic showing the basics of a photoelectron emerging from the surface of a solid. The potential step as the electron transits from the solid to the vacuum at the surface reduces the perpendicular component of its momentum.

shown in Fig. 5. Figure 6 exhibits the refraction of the electron as it transits the surface due to the potential step normal to the surface. To understand this step we can write the total kinetic energy of the electron in the vacuum, external to the solid as

$$E_{\mathrm{kin}} = (\hbar^2/2m)(k^2\|_{\mathrm{o}} + k^2\perp_{\mathrm{o}}), \tag{5}$$

where $k\|_{\mathrm{o}}$ and $k\perp_{\mathrm{o}}$ are the outside or vacuum parallel and perpendicular components of the electron wavevector. While the perpendicular component of the momentum changes as the electron transits from the sample into the vacuum due to the perpendicular potential step, the parallel component is conserved, modulo a surface reciprocal lattice vector.

From Fig. 6 we can relate the parallel component of the electron momentum both in the bulk and at the surface to the measured kinetic energy and the emitted angle θ_{o} of the electron by the following simple relation

$$k\| = [(2m/\hbar^2)(E_{\mathrm{kin}})]^{1/2}\sin\theta_{\mathrm{o}}, \tag{6}$$

where

$$\mathbf{k}\|_{\mathrm{o}} = \mathbf{k}\|_{\mathrm{in}} + \mathbf{G}_{\mathrm{surface}} \tag{7}$$

and $\mathbf{G}_{\text{surface}}$ is a surface reciprocal lattice vector. These relations permit the evaluation of the parallel component of the electron momentum within the bulk as well as for the surface, where only $k\|$ is of concern owing to the two-dimensional nature of the surface. Surface states are uniquely identifiable through their lack of dispersion normal to the surface. While the binding energy of a bulk state will be observed to shift as the photon energy is varied (see Fig. 5), the binding energy of a surface state will remain fixed. Again, binding energy is related to the Fermi level of the system. In addition, a surface state will be observable over essentially all photon energies; photoelectrons that derive from a surface state couple directly to plane wave vacuum states in a single step. Intensities may change as there are resonant enhancements due to contributions from the bulk band structure. In addition, surface states can be identified by their sensitivity to contamination from water or other gaseous species.

Bulk band structure evaluation is somewhat more involved since the three-dimensional bulk is observed through a two-dimensional surface with the addition of the aforementioned potential step as the electron transits from the bulk to the vacuum. A convenient and experimentally confirmed approach to solve this issue is to first assume that the final state band dispersion can be well approximated by a free electron-like band offset by an inner potential, E_{o} referenced to the valence band maximum. This level is shown in Fig. 5. As UV photons carry near zero momentum resulting in vertical transitions within the reduced Brillouin zone, the initial and final k values are the same. The final energy $E_f(\mathbf{k})$ is the sum of the initial state energy $E_i(k) + h\nu$. The final electron energy is

$$E_f(\mathbf{k}) = E_i(\mathbf{k}) + h\nu = (\hbar^2/2m)(k^2\| + k^2\perp) + E_{\text{o}} = E_{\text{kin}} + e\varphi, \qquad (8)$$

where E_{kin} is the electron kinetic energy and $e\varphi$ is the photothreshold for the material (vbm to vacuum level). E_{o} is negative as it is measured relative to the vbm and referring to Fig. 5, where $V_{\text{o}} = e\varphi - E_{\text{o}}$ we can solve for $k\perp$ and find

$$k\perp = [(2m/\hbar^2)(E_{\text{kin}}\cos^2\theta + V_{\text{o}})]^{1/2}. \qquad (9)$$

Once the inner potential is known the bulk band structure can be mapped and a particular feature in the photoelectron spectrum can be identified [19].

The above set of equations allow for the full characterization of the electronic band structure at the surface and in the bulk. While band structure of many crystalline materials such as group III–V's and metals throughout the periodic table have been studied, new materials are constantly being introduced for their unique properties and potential applications in devices ranging from semiconductor integrated circuits to photovoltaics (PVs), sensors, spintronic devices, and more.

Electron detectors for ARUPS must, by definition, subtend a narrow angular region, in order to extract the momentum information carried by the emitted electron. In its most simple form, the emitted electron is collected with an electron detector fitted with an aperture that subtends an appropriately narrow angle, $\delta\Theta$ for electrons emitted at an angle Θ, relative to the surface normal as shown in Fig. 4.

As a simple example, consider the two-dimensional band structure of a surface state. If we wish to determine the angular resolution we need for our detector in order to achieve a specific resolution within the Brillouin zone, we calculate the first derivative of $k\|$ which yields, for an actual angular resolution $\delta\Theta$.

$$\delta k\| = [(2m/\hbar^2)(E_{\text{kin}})]^{1/2} \cos\Theta\delta\Theta. \tag{10}$$

For electrons emitted near the surface normal such that $\Theta = 0^0$, and for an electron kinetic energy of $E_{\text{kin}} = 5\,\text{eV}$, an angular resolution of $\pm 3^0$ results in a momentum resolution $\delta k\| = \pm 0.06\text{Å}^{-1}$ corresponding to 5% of the Si Brillouin zone along the Γ-X symmetry line.

4.6 Angle-Integrated UPS

While we have been previously discussing ARUPS, most materials, particularly those of industrial interest are much more complicated. Materials that are not single crystal in nature include all forms of small grain polycrystalline materials (grain size $<$ 100 microns), semiconductor and metal oxides, amorphous materials (amorphous hydrogenated Si, for example), conductive glasses, biological materials and more. For these materials, it is more effective to carry out angle integrated UPS where the photoelectron spectrum reflects the actual density of states of the system as a function of energy. Figure 7 shows a cross-sectional scanning electron micrograph (SEM) of a polycrystalline film of $Cu_2ZnSn(S_xSe_{1-x})_4$ also known as CZTSSe, a promising earth abundant photovoltaic (PV) absorber. As can be seen multiple individual crystalline grains of CZTSSe are oriented in a relatively random pattern. As a result, it is more useful to perform photoemission studies of this material with an angle integrated detector. Further details of these studies are discussed in Chapter 6.

Again, relating UPS spectra from polycrystalline materials to the system Fermi level provides an energy scale zero. Angle integrating detectors exist commercially in a number of forms such as hemispherical and cylindrical

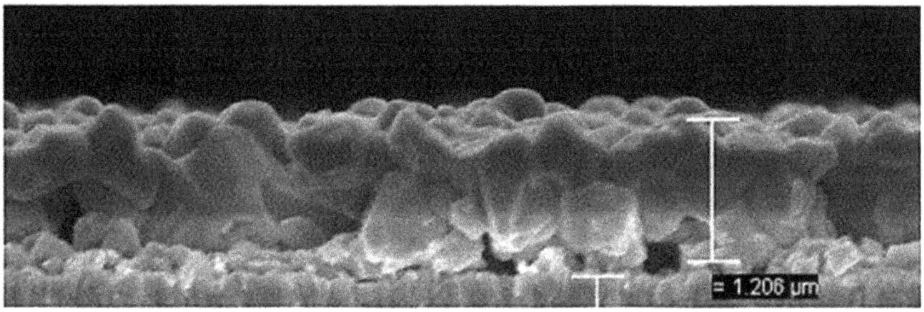

Fig. 7. Cross-sectional SEM of CZTS, Se showing the polycrystalline nature of the film and the varying orientations of the crystal grains.

mirror electrostatic analyzers. Many of these modern analyzers use one- or two-dimensional detector arrays so that many angles can be accessed simultaneously to increase data throughput and minimize data collection times.

4.7 Angle-Integrated Parabolic Mirror Time-of-Flight Electron Analyzer

One very useful approach utilizes time-of-flight (TOF) electron detection and analysis, particularly well-suited to the use of femtosecond lasers. One flavor of this type of analyzer that has been used successfully is a parabolic mirror TOF analyzer [20], versions of which are shown in Fig. 8. In this detection scheme, a section of a paraboloid is employed. A paraboloid is an imaging surface with minimal aberrations and as a result an electrostatic detector based upon this solid conic section will also exhibit minimal aberrations. In this case, an imaging detector implies that all path lengths originating from the focus of the paraboloid are equal. Figure 8 shows the parabolic section mirror analyzer.

In the employment of this analyzer, the energetic light used for UPS is focused on the sample. The light spot on the sample is then the focal point of the parabolic mirror. Electrons are emitted over 2π steradians and those electrons with angles subtended by the mirror are then imaged at infinity with path length and emitted angle conserved. Electron drift times [20] are measured with accurate timing electronics and the energies of the electrons are calculated with the simple relation

$$E = (1/2M_e)L_{\text{drift}}^2/T_{\text{drift}}^2, \tag{11}$$

where M_e is the electron mass, and L_{drift} and T_{drift} are the electron drift lengths and times. Modern timing electronics can measure the drift times with $\sim$100 ps resolution. Energy resolution is energy dependent for this type

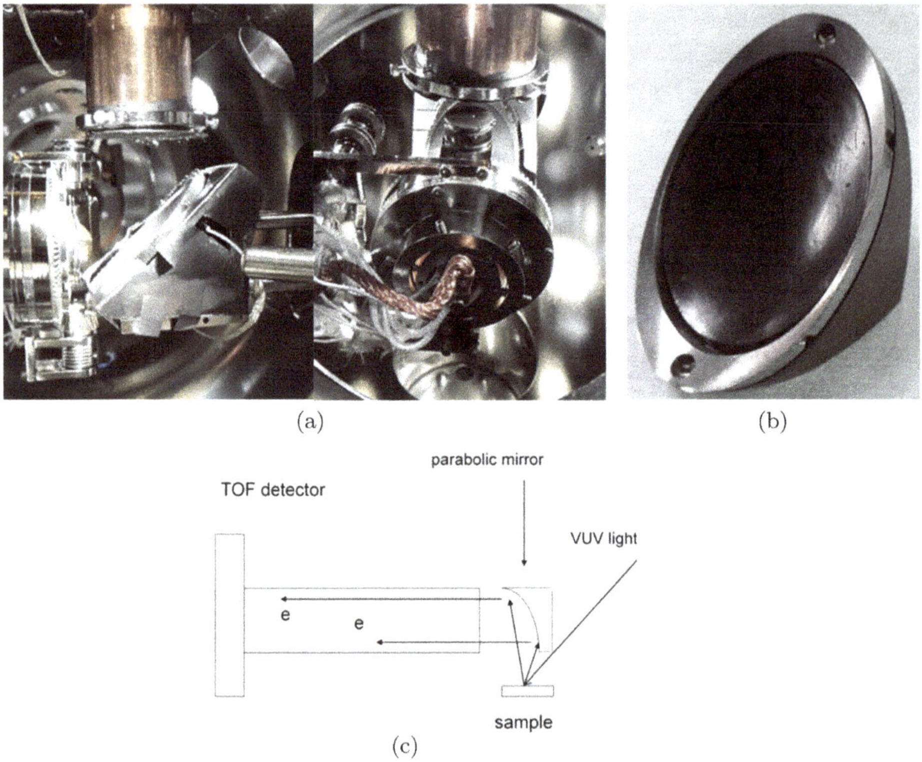

Fig. 8. (a) Left: side view of sample manipulator, parabolic mirror, and drift tube. The vacuum UV laser light emerges from horizontal re-entrant tube on right and is focused on the sample mounted on a multiaxis manipulator on left. Emergent electrons are collected by the mirror, their trajectories parallelized and directed vertically into the field-free drift tube. Right: top view of sample manipulator, parabolic mirror, and Cu drift tube. On the right is re-entrant tube with small aperture. (b) parabolic mirror machined from a Cu block, coated with graphite. The non-magnetic stainless ring holds a high transparency metal mesh (not shown) that provides a ground plane confining the electrostatic field between the mirror and mesh. (c) schematic of detection system.

of detector with the resolution given by

$$\Delta E = 2E\Delta T/T_{\text{drift}}, \tag{12}$$

where E is the electron energy, ΔT is the timing resolution, in this case 100 ps and T_{drift} depends on the length of the TOF region and electron energy. As a concrete example, it takes a 10 eV electron ~260 ns to drift 100 cm (a reasonable length of field-free drift). With our timing resolution of 0.1 ns, we achieve an energy resolution of ~4 meV. Actual resolutions are typically much larger than this due to jitter in the timing electronics that can broaden ΔT to >0.5 ns. Additionally, stray fields, aberrations associated with extended focal regions and imperfections in machined surfaces also conspire to degrade energy resolution. Nonetheless, very high energy resolution can be achieved with this approach.

4.8 Pump-Probe Photoelectron Spectroscopy

Why is photoelectron spectroscopy so important and useful in the study of the electronic properties of condensed matter? Arguably, the answer lies in the fact that it is one of the few experimental techniques that interrogates the electronic states directly. Optical spectroscopies investigate changes in reflectivity or transmission through a material or stimulated photoluminescence (PL) where excited electron-hole pairs recombine following absorption of a photon. Since this occurs over the absorption length of the light which can be hundreds of nanometers or more, integrated bulk information is revealed. As such the complementary nature of these techniques enables deep characterization of the electronic properties of the material and its surfaces and interfaces.

The advent of ultrafast laser sources starting in the 1970s and improvements over the years to present day created an entirely new experimental vista. In addition to characterizing and understanding the static electronic properties of a material, it was now possible to plumb the depths of the ultrafast electron dynamics. Pump-probe photoelectron spectroscopy has its origins in the early work [21] that used two photons, typically a red light pulse $\hbar\omega \sim 2.3\,\text{eV}$ and a blue photon $\hbar\omega \sim 4.6\,\text{eV}$ such that the sum of the two photons promoted electrons to states above the vacuum level. The blue photons were generated by frequency doubling of the fundamental red light. An excellent review by Petek and Ogawa [22] details the physics and applications of femtosecond 2-photon photoemission.

An excellent example of the utility of this approach was the use of two photon photoemission to measure the cooling rate of hot electrons in a thin Au film thereby establishing the timescale for the electron–phonon interaction in metals [22, 23] (Fig. 9). Such time scales are exceptionally important in the area of ultrafast laser ablation for machining and other applications as discussed in Chapter 5.

In this particular two photon photoemission experiment, femtosecond pulses of 800 nm light were directed onto the surface of a Au foil [23]. Upon absorption of the light pulse the excited electrons rapidly thermalize to form a hot electron population in the Au conduction band. Cooling of the electron population occurs by emission of phonons, i.e. lattice vibrations, that transfer the excess energy of the electron population to the lattice. This

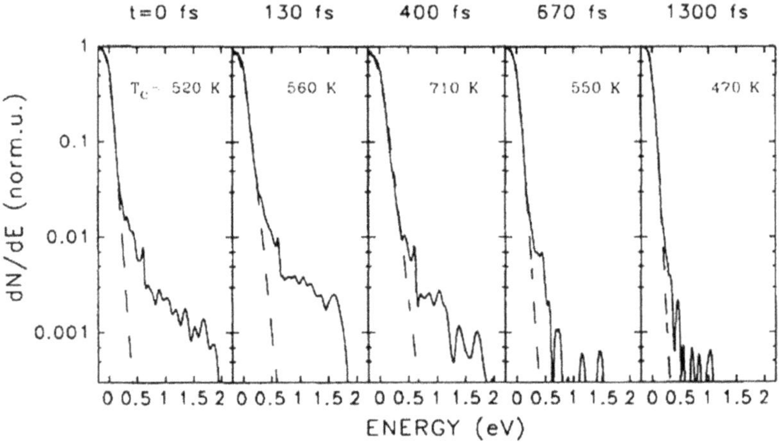

Fig. 9. Results of 2-photon photoemission showing the cooling of the photoexcited electron population within Au as a function of delay time. A cooling rate of ~1 ps was determined in this experiment [23, 24].

cools the electron population while it heats the lattice itself. This is the fundamental nature of laser heating and melting of the metal. As can be seen in Fig. 9, the electrons are excited to states well above the Au Fermi level as evidenced by extended foot in the photoelectron emission spectrum. By varying the delay between the probe UV pulse relative to the IR excitation pulse the retreat of the excited electron population back to its equilibrium distribution at room temperature can be followed, and a cooling rate can be extracted.

While two-photon photoemission is an effective method for studying electronic states near the Fermi level, the low photoemission photon energy limits the range of states that can be studied. There are also issues of background subtraction at low electron kinetic energies. This background is a consequence of utilizing short pulses of light to pump the sample. Multiphoton absorption from the pump light can also result in the photoemission of electrons that creates a background relative to the IR + UV process. To control this, relatively weak pulses of IR must be used, particularly as the excitation and probe pulse duration becomes shorter. This then limits how dense an excited electron population can be created.

In order to overcome these issues, the probe pulse can be upconverted to a much higher photon energy via high-harmonic generation [16, 17]. The details of this were discussed in Chapter 3. In essence, an intense pulse of laser light, typically the output of a Ti:sapphire laser operating at 800 nm, is focused into a pulse of a rare gas such as He, Ne, Ar, Kr, or Xe. Ar is an inexpensive welding gas, so it is typically employed for applications in

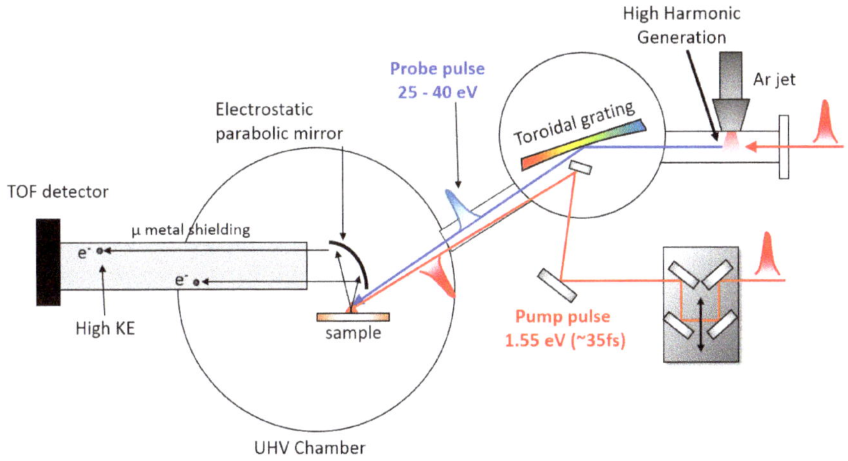

Fig. 10. Schematic of a fs-UPS system. Individual pulses produced from a femtosecond, amplified laser system are split into 2 synchronized pulses. The pump pulse (red) is used to photoexcite the sample under study while the probe pulse is converted to the XUV by focusing into a pulse of Ar. A grating selects the harmonic of interest and focuses onto the sample in the vacuum chamber. A TOF electron spectrometer measures the kinetic energy of the photoemitted electrons.

harmonic conversion to photon energies below ~100 eV. A schematic setup of a femtosecond vacuum UV photoelectron spectroscopy system [25] is shown in Fig. 10.

In this arrangement, an amplified femtosecond laser system, typically operating at a repetition rate of 1 kHz or higher produces 800 nm pulses of light with 1–10 mJ of energy per pulse. The pulse can be split into two synchronized pulses by a simple beam splitter. One of the 800 nm pulses, called the pump, is then directed through a delay line that allows control of the arrival time of the pulse at the sample. This pulse is typically weaker than the probe pulse by virtue of the ratio of the reflected to transmitted light engineered into the beam splitter.

The more intense pulse, the probe, is directed into a vacuum beam line that possesses a mechanism for introducing a puff of rare gas into the path of the focused probe light. One approach to introduce the gas is to employ a piezoelectric pulsed valve that opens briefly when a short, high voltage pulse is applied. At the focus of this light, the intensity may be as high as 10^{14}–10^{15} W/cm^2. At these intensities the rare gas atom, typically Ar, is ionized by the intense laser field. The ionized electron is captured by the oscillating light field and is redirected back toward the ionized Ar nucleus where it can be recaptured [16]. The electron can acquire energy from the laser light field and that energy is converted into light during recombination with the nucleus.

This excess energy results in the production of high frequency photons whose energies are odd multiples of the fundamental 800 nm driving laser. Since significant emission at even multiples is forbidden by the inversion symmetry of free space, only odd multiples of the driving laser frequency are found. The details of high-harmonic upconversion of laser light were discussed in detail in Chapter 3.

Other methods for generating high harmonics were discussed in Chapter 3. One method for efficient generation of harmonics involves the use of a capillary waveguide [26–28]. In this approach, an intense pulse of femtosecond light is focused into the front end of a glass capillary filled with a rare gas such as Ar, Xe, Kr, He, or H_2. A guided mode travels down the length of the waveguide. Careful control of gas pressure in the capillary allows for phase matching of the driving 800 nm pulse and the generated harmonics. Propagation of the pulses within the plasma formed in the capillary is a key consideration. The width of the capillary can be modulated during its fabrication to contribute to the phase matching. Phase matching then results in the enhancement of particular harmonics of interest optimized with gas pressure. In this manner, generated harmonic output can be increased substantially.

Now that high energy photons are routinely available from a table-top amplified laser source an entirely new set of experiments can be carried out that interrogate the state of the system under excitation. As shown in Fig. 11 an intense pump pulse (shown by the vertical block arrow) promotes electrons from the region in k-space near the top of the valence band (at and near the Brillouin zone center in this case) to unoccupied states in the conduction band. Low energy photons carry near-zero momentum so that the transition is vertical. While the case shown in Fig. 11 is for a direct gap material where both vbm and conduction band minimum are both located at Γ, the Brillouin zone center this discussion can be generalized to any material, direct gap or indirect. Electrons excited into the conduction band states at or near the surface can equilibrate via electron–electron scattering, and then lose energy via electron–phonon scattering. For metal systems, this can occur on the order of a picosecond while for semiconductors the scattering time is on the order of a few hundred femtoseconds. At longer timescales, typically greater than nanoseconds, electrons and holes will recombine radiatively resulting in the emission of photons [6] or non-radiatively through defect mediated processes.

Since photoelectron spectroscopy is inherently surface sensitive as has been described earlier in the chapter, the density of excited electrons at the surface will decay via diffusion into the interior of the material. Diffusion is

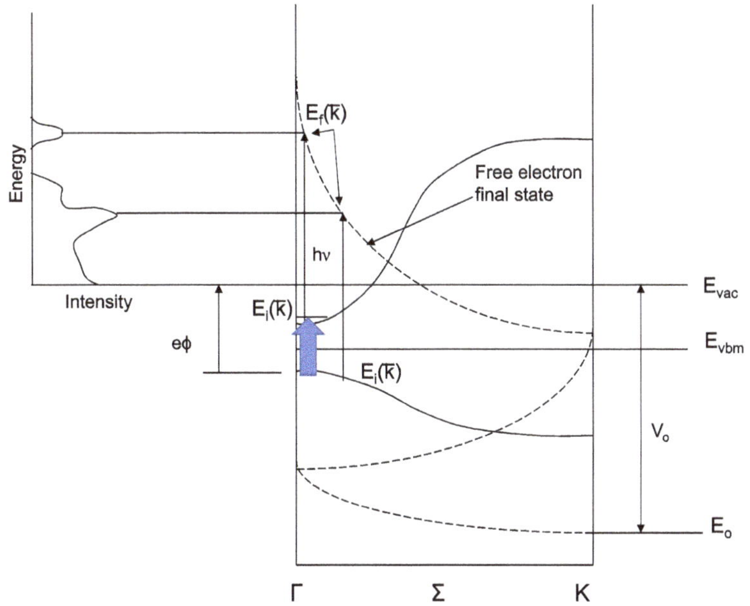

Fig. 11. Similar to Fig. 5 previously in this chapter, this figure adds the case where femtosecond pulses are used to first excite electrons from the valence band into the conduction band (shown by vertical block arrow) followed by a time synchronized pulse that photoemits both valence and transiently excited electrons to states above the vacuum level. The resultant photoemission spectrum is shown on the left.

driven by the initial population gradient set up by the absorption of light given by $I = I_o e^{-(\alpha z)}$ as described earlier (Eq. (2)).

The exponential decay of light into the absorbing material indicates that the initial excited electron distribution will decay exponentially into the material as well. This distribution results in a gradient that drives diffusion of electrons away from the surface into the interior of the material and must be accounted for in the temporal evolution of the excited electrons [29].

Additionally, defects at the surface can lead to non-radiative electron-hole recombination. This is typically referred to as a surface recombination velocity and is a critical consideration at interfaces where high recombination velocities can rob devices of their performance. In semiconductor devices, trapping of charge at interface states can impact the capacitance of an MOS (metal-oxide-semiconductor) device. High capacitance associated with interface states can increase device switching times giving rise to poor device performance. Another example where defects at interfaces and their dynamics are critical is in the performance of PV devices. High surface recombination velocities can reduce short circuit currents by limiting the collection of light-induced carriers and limit open circuit voltages achieved in PV devices.

4.9 Pump-Probe Photoelectron Spectroscopy with High Harmonics — Examples

A number of experiments exemplifying what can be learned regarding electron dynamics at and near surfaces are described below. Early experiments on cleaved III–V surfaces include InP [30] and GaAs [29] revealed electron diffusion, surface intervalley scattering, surface recombination, and interface states.

Figure 12 displays results of sub-picosecond pump/probe UPS experiments on the cleaved GaAs (110) surface [29]. Electrons were photoexcited from the occupied GaAs valence bands into the empty conduction bands with a sub-picosecond pulse of light. The photoexcited electrons relaxed into the bottom of a surface resonance state at the Brillouin zone center $\bar{\Gamma}$ and were captured in the UPS spectrum as a peak at 1.4 eV above the vbm (Fig. 12 top spectrum). Since excited electrons will seek their lowest accessible energy state, they also scatter into a surface state at the Brillouin zone edge at $\bar{X}$. This empty surface state was theoretically predicted [31], but observed for the first time in this experiment. Electrons at this point in the surface Brillouin zone will carry momentum lateral to the surface.

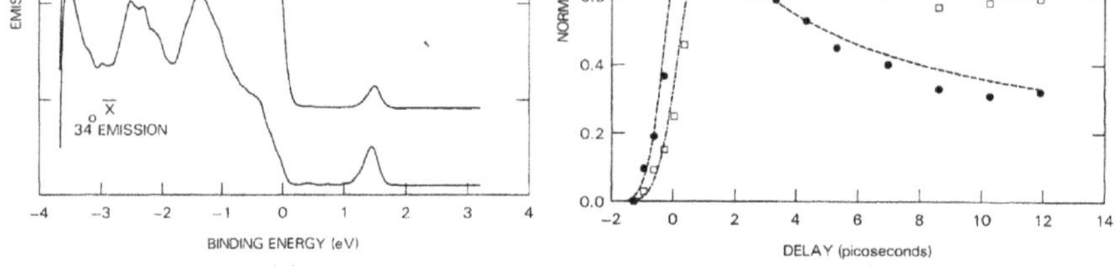

(a) (b)

Fig. 12. (a) Pump/probe photoemission spectra collected in normal emission (top spectrum) and 34° emission corresponding to the surface Brillouin zone edge. (b) Time dependence of photoexcited electrons from normal and zone edge emission. The electron–phonon scattering time is responsible for the delay in the zone edge emission relative to zone center.

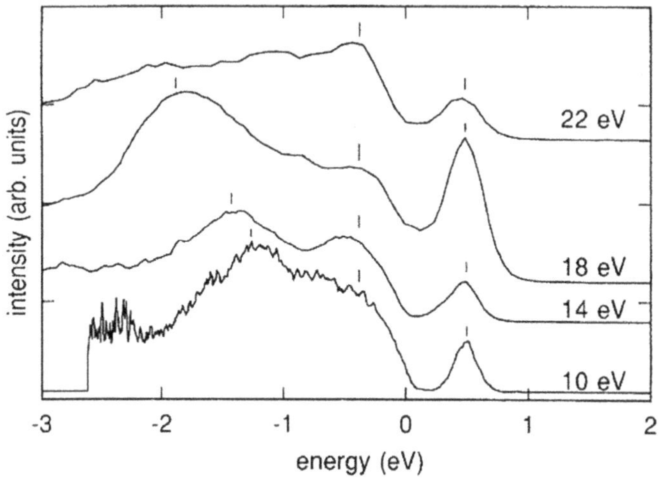

Fig. 13. Series of pump/probe photoelectron spectra of As-terminated Ge(111) at 4 different XUV energies. The peak at 0.4 eV derives from electrons photoexcited into an arsenic derived empty state at the surface zone center. Its intensity peaks at 18 eV photon energy indicating a resonance in the photoemission cross-section at this energy.

Using Eq. (6) delineated earlier in this chapter, we can calculate the angle of emission for electrons that pool at the $\bar{X}$ location, the surface zone edge, whose wavevector is 0.786 $\dot{A}^{-1}$. For electrons that scatter to this point in the Brillouin zone, the calculated emission angle is 34° relative to the surface normal (bottom spectrum). Figure 12(b) displays the time evolution of the electron populations at both the zone center $\bar{\Gamma}$ and edge $\bar{X}$; the delay in the $\bar{X}$ population of 400 fs corresponds to the scattering time of electrons to this location from the zone center. Hence, this experiment represents the first example of the surface analog of bulk intervalley scattering in GaAs that relates to the original work on the Gunn effect in III–V materials [32].

A second example [33] is shown in Fig. 13 where the arsenic terminated Ge(111), As–Ge(111), surface displays a normally empty antibonding state 0.4 eV above the vbm.

The arsenic terminated Ge(111) surface is an ideal, unreconstructed atomic surface; this surface is particularly inert and exists in ultrahigh vacuum for weeks without change. In this experiment, UV photon tunability was demonstrated through the use of a grating to select a series of harmonics of increasing energy. At 18 eV photon energy, the 0.4 eV peak emission intensity increases dramatically indicating a resonance in the photoemission process.

Organic light emitters have also been studied with pump/probe UPS [34]. Figure 14 displays the normally occupied and transiently excited states

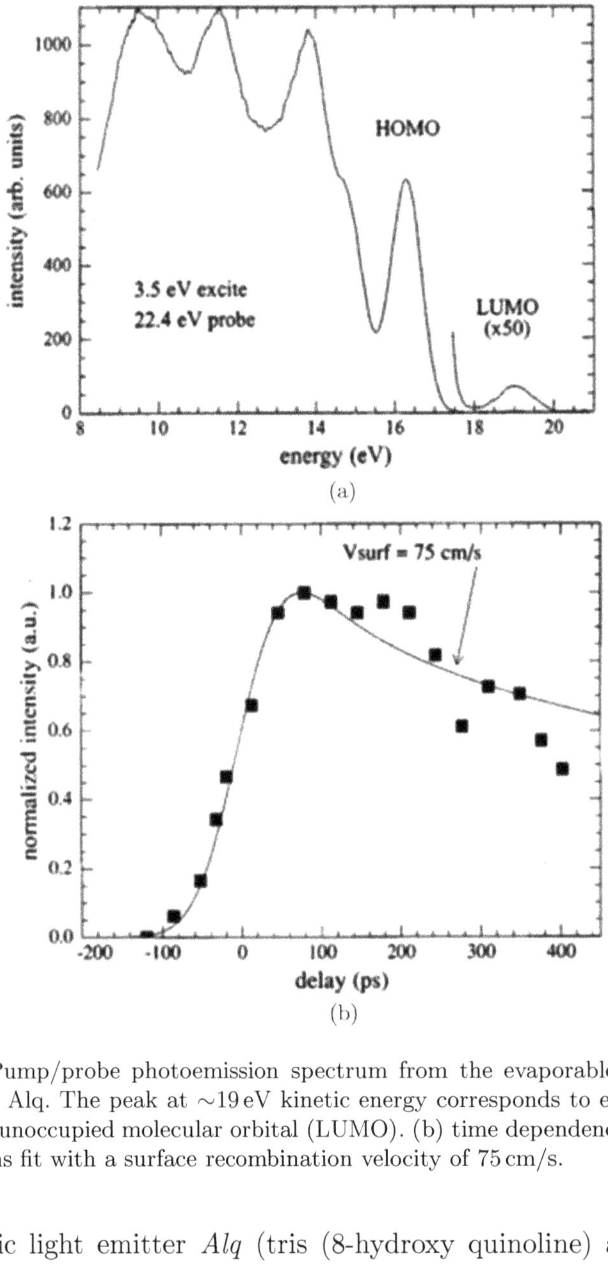

Fig. 14. (a) Pump/probe photoemission spectrum from the evaporable light emitting molecular layer Alq. The peak at ~19 eV kinetic energy corresponds to electrons excited into the lowest unoccupied molecular orbital (LUMO). (b) time dependence of the excited LUMO electrons fit with a surface recombination velocity of 75 cm/s.

of the organic light emitter *Alq* (tris (8-hydroxy quinoline) aluminum), a key evaporable material used in organic light emitting displays (OLED). OLED displays have become commercially available, employed in both cell phone displays and in flat-panel televisions. The time dependence of the excited electrons is shown in the lower panel of Fig. 14 and the highest occupied (HOMO) to lowest unoccupied molecular orbital (LUMO) gap can be accurately determined.

Metal chalcogenides (WSe_2 and $MoSe_2$) are interesting layered semiconductors. Much like mica and graphene, these layered dichalcogenides form two-dimensional layers, with the metal atom (W or Mo in this case) sandwiched between the chalcogen atom (Se or S). The layers are bound via van der Waals forces in the vertical dimension. The layers are easily separated in a cleaving or exfoliation process to reveal a fresh surface that is inert and exhibits long-term stability in vacuum. The highly anisotropic nature of the material suggests that the electronic transport properties will exhibit anisotropies as well. Fs-UPS experiments [35] carried out on $MoSe_2$ and WSe_2 provide experimental evidence of such ultrafast anisotropic transport. In Fig. 15, the pump/probe UPS spectrum from $MoSe_2$ is displayed showing

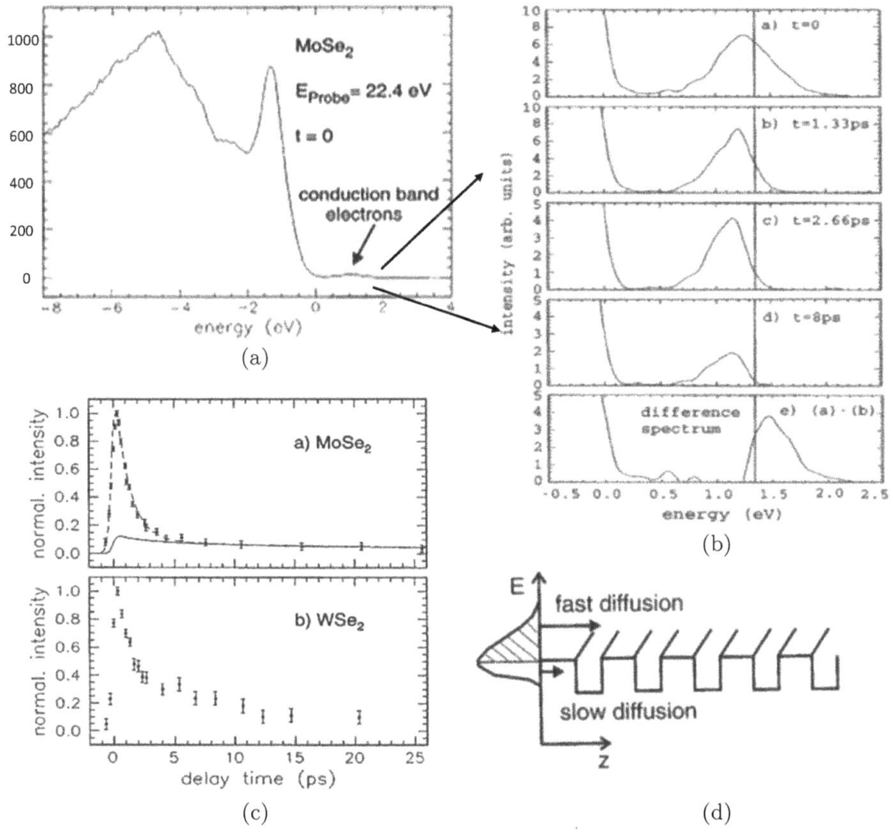

Fig. 15. (a) pump/probe photoemission spectrum of cleaved $MoSe_2$ showing the excited conduction band electrons at 1 eV. (b) focusing on the excited states showing the time evolution of the 1.1 eV peak. (c) decay rates of electrons excited into the conduction bands of $MoSe_2$ and WSe_2 showing a rapid decay followed by a second much slower decay. (d) model explaining fast and slow diffusion dynamics.

the valence states, and the appearance of a transient peak at 1.1 eV, the band gap of $MoSe_2$. Detailed analysis of the transiently populated conduction band peak shows that at the time of excitation, a hot electron population exists, as evidenced by the high energy tail shown in Fig. 15(b). At 1.3 ps delay, the high energy tail disappears while the main peak at energies below the vertical line has not changed in intensity. At later times the entire excited state emission signal decays. The difference spectrum in Fig. 15 was generated by subtracting the $t = 0$ signal from the $t = 1.33$ ps signal. As can be seen, the difference is concentrated in the excited high energy electrons. The initial loss of the high energy side of the peak suggests a rapid diffusion of electrons into the interior of the material with a subsequent slow decay for the lower energy electrons. This is confirmed by noting the excited state electron decay dynamics shown in Fig. 15(c). For both $MoSe_2$ and WSe_2 the electron dynamics can be modeled by an exponential decay with a time constant of 1.1 ps to account for the initial rapid loss of intensity (dashed curve) and slower dynamics modeled by solving the diffusion equation with a diffusion coefficient D $= 1\,\text{cm}^2/\text{s}$. The schematic in Fig. 15(d) suggests

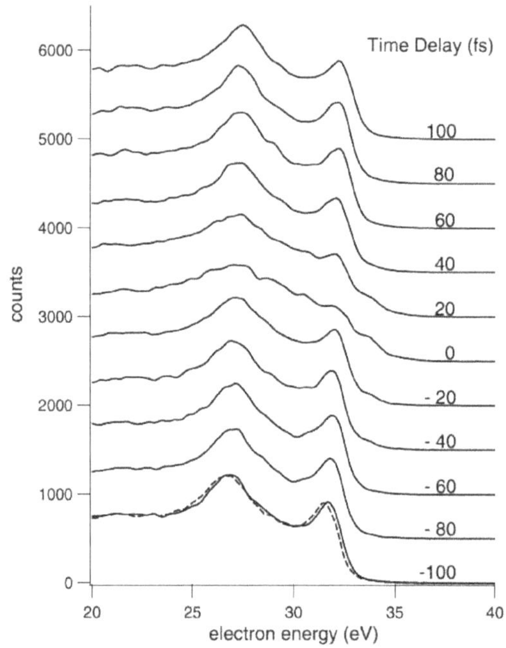

Fig. 16. Time resolved photoelectron spectra from Pt(111). At and near $t = 0$ an additional feature is observed above the Fermi level (here at $\sim$34 eV) derived from the overlap of a 1.6 eV pump pulse and the 27th harmonic-42 eV — from an amplified Ti:sapphire laser system [40].

that the energetic electrons rapidly transit into the interior of the material while the lower energy electrons experience a greater degree of confinement within the layers that inhibits diffusion into the material.

While the experiments described above utilized pulses in the picosecond to sub-picosecond range, the development of shorter pulsed lasers down to the few tens of femtoseconds and then into the attosecond regime drove significant progress in the types of time and angle UPS experiments that were carried out in materials [36–39]. Experiments utilizing ultrashort pulses include laser assisted photoelectric effect (LAPE) [40], where cross-correlation of a high energy and low energy photons results in side-bands that can be observed in photoemission spectra. An example is shown in Fig. 16.

More recent work [41] on charge-density-wave insulators (1T-TaS_2, 1*T*-$TiSe_2$ and Rb:1*T*-TaS_2) utilized time and angle resolved photoemission to measure the melting times of the electronic order parameter in these unusual insulators. In Fig. 17, a unique spectrometer with a momentum space view

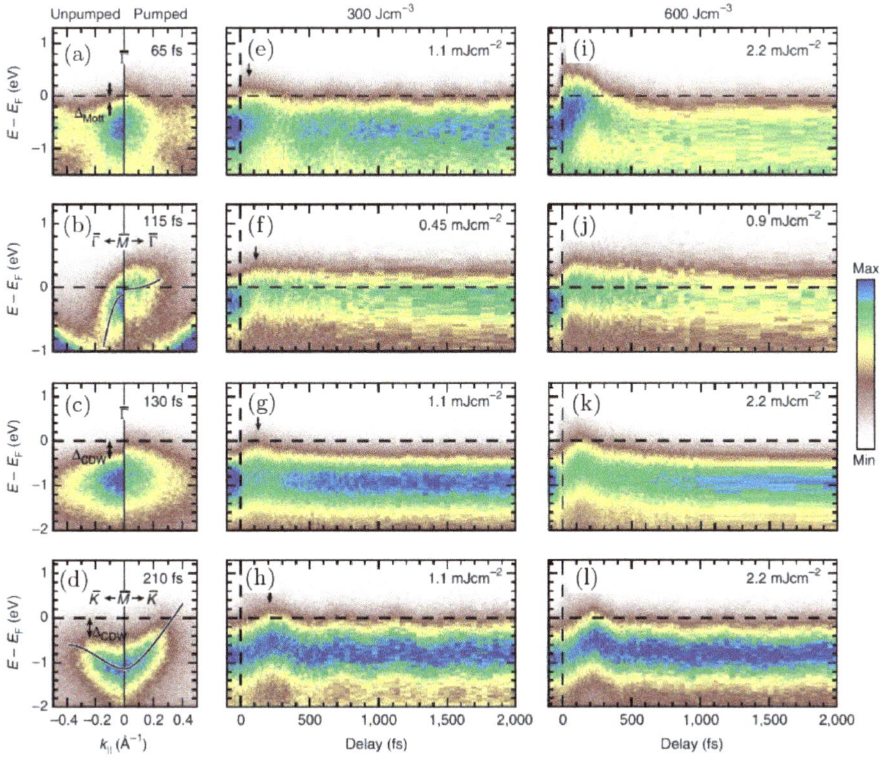

Fig. 17. Time and angle resolved photoemission from (a–d) 1T-TaS_2, 1T-$TaSe_2$, Rb:1T-TaS_2 at various points within the Brillouin zone, for different pump fluences [41].

of the electronic structure shows the excited electron dynamics of these insulators at specific points in k-space for two different pump intensities. These investigations were used to analyze the melting times of the electronic order parameter to identify the dominant interaction. The mechanisms involved were hotly contested and fs-UPS investigations were key to resolving this controversy.

Clearly, such ultrafast, few femtosecond photoemission experiments are extraordinarily useful in delineating the fundamental electronic and structural processes that occur at extremely short times.

As a final example we describe a method for studying materials at the nanoscale with fs-UPS. Figure 18 shows a schematic view of a single nanowire

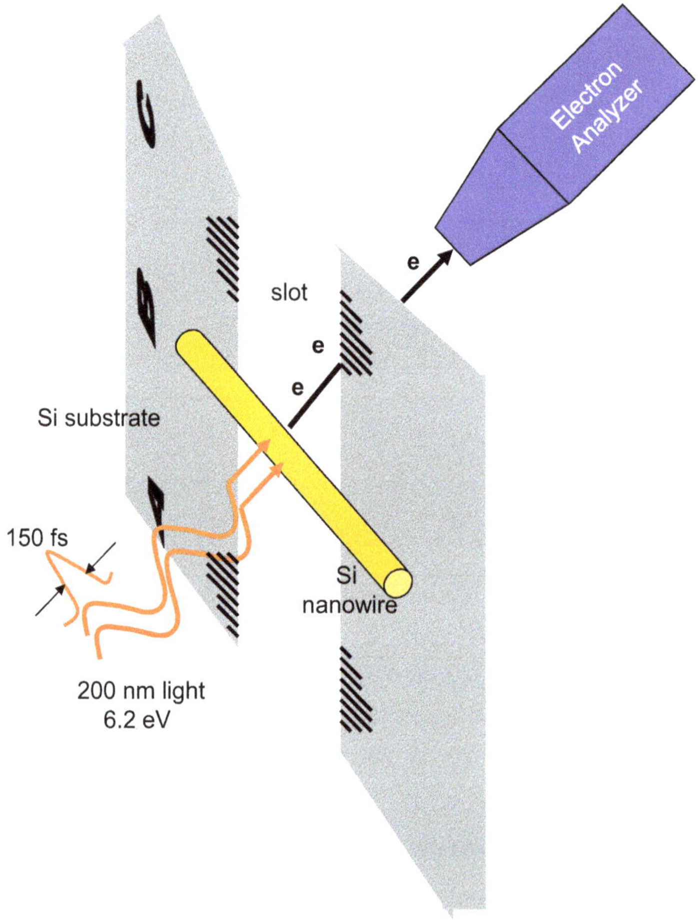

Fig. 18. Schematic of single NW photoemission where an individual NW is suspended across a gap. 6.2 eV photons impinge on the front of the NW while photoelectrons emerge from the back surface and are collected by the electron detector.

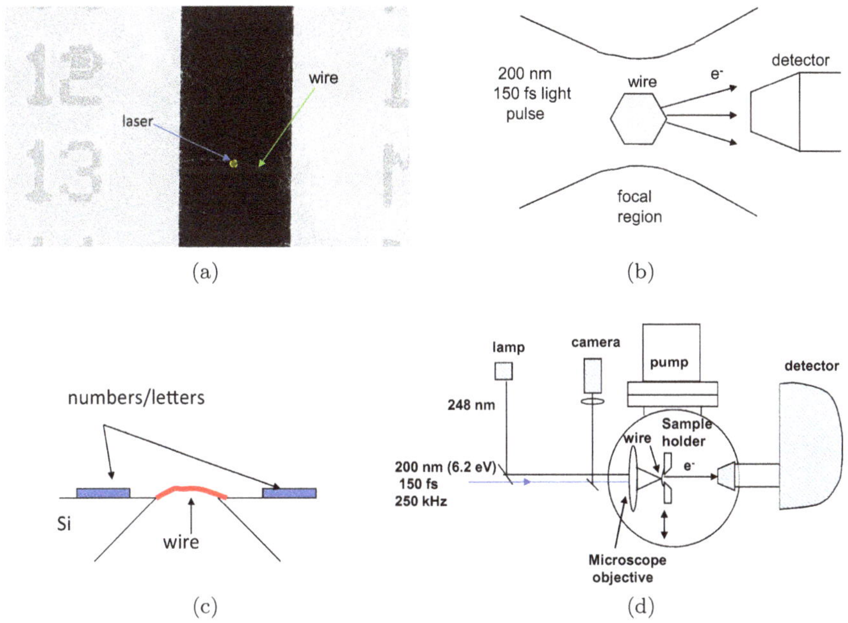

Fig. 19. (a) SEM showing an individual Si NW suspended across the Si wafer gap, suitably dressed with markings for location with an optical microscope. (b) schematic of Si NW at focus of 6.2 eV light pulse, (c) side view of Si NW crossing gap, (d) schematic showing vacuum apparatus, light delivery, monitoring and location of electrostatic electron analyzer.

NW stretched across a gap etched into a Si wafer dressed with fiducials for locating the NW [42]. In these experiments, Si, Ge, and ZnO NWs are grown [43, 44] on pre-patterned Si wafers by the vapor–liquid–solid method utilizing Au nanoparticles to nucleate growth. NWs were randomly distributed across the wafer and sporadically found to span a gap in the Si wafer. Those wires spanning the gap were identified by searching with SEM; an example is shown in Fig. 19.

UV light at 200 nm (6.2 eV) is generated as the fourth harmonic of 800 nm, 150 fs pulses of light from an amplified Ti:sapphire laser utilizing beta-barium borate (BBO) nonlinear crystals for frequency doubling and summing. The intensity of UV light at 6.2 eV drops to 1/e of its initial value within ~10 nm of the surface. The inelastic mean-free-path for electrons with an energy 6 eV above the Fermi level is ~6–10 nm. Thus, for Si NWs whose diameters are on the order of several 10 s of nm, photoexcitation as shown in the geometry of Fig. 18 will produce photoelectrons that emerge from the NW surface and interior and can be collected by an electron detector positioned behind the wire.

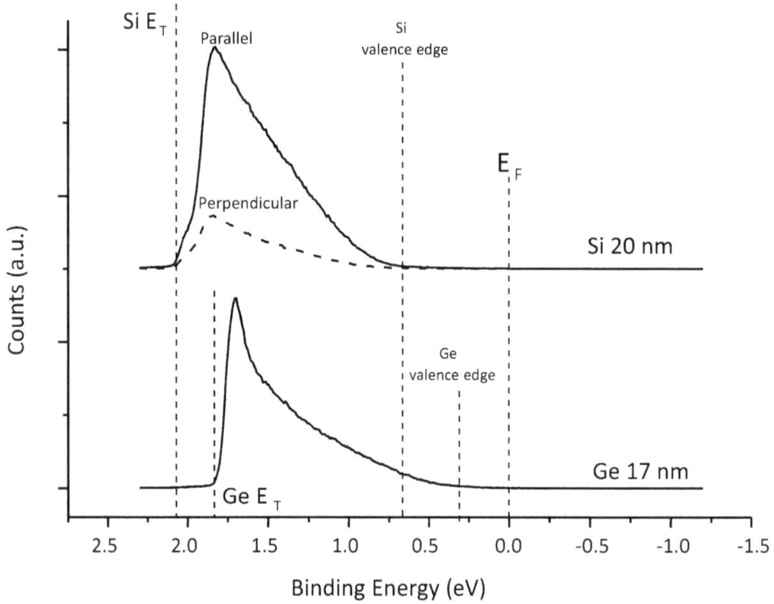

Fig. 20. Photoemission spectra from a 20 nm diameter Si NW and 17 nm Ge NW. E_T is the threshold emission energy. For Si the anisotropic emission intensity for 6.2 eV light polarized parallel (solid spectrum) and perpendicular (dashed spectrum) to the long axis of the NW is shown.

Focusing of the 6.2 eV light is achieved through the use of a reflective Schwarzchild microscope objective that also images the Si substrate. Figure 20 shows an example of the collected photoelectron spectra from a 20 nm Si and 17 nm Ge NW. Light polarized parallel or perpendicular to the long axis of the NW showed considerable anisotropy consistent with Mie scattering physics [42].

Utilizing this experimental approach to directly studying semiconductor NWs, Al doping of individual Si NWs was studied [45]. Figure 21 shows the utility of direct UPS on Al-doped Si NWs. In this experiment, Si NWs seeded with Au nanoparticles are compared with those seeded with Al. In contrast to Au, Al is soluble in Si and hence dopes the NWs during growth; Al is a *p*-type dopant in Si. As can be seen in Fig. 21, Al incorporation at a growth temperature of 460°C moves the Fermi level, E_F, toward the vbm of the Si NW due to the Al incorporation. At 490°C E_F is shifted closer to the vbm. Following a 10 s anneal at 900°C there is a dramatic shift of E_F toward the vbm, consistent with increased activation of the Al present in the NW. Figure 21(b) shows the activated concentration of Al dopant in the NW. Note

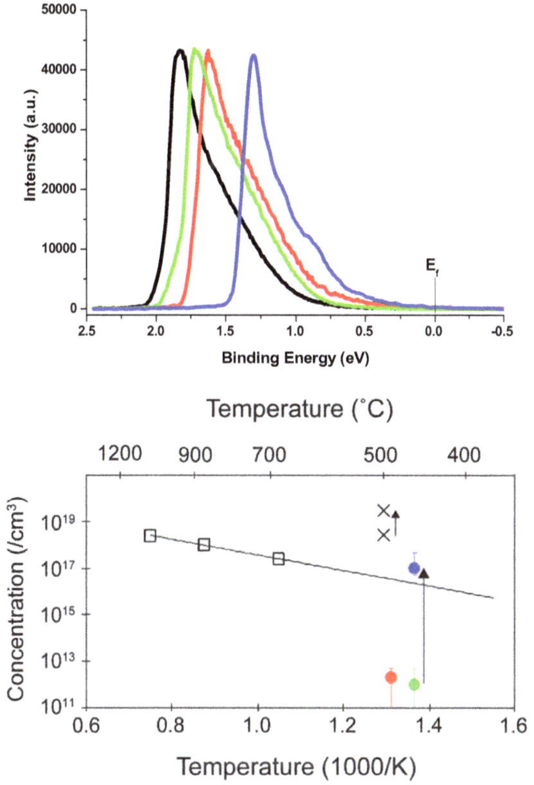

Fig. 21. (a) series of NW photoelectron spectra. Black: NW grown with Au nanoparticles, Green: Al seeded NW grown at 460°C, Red: Al seeded NW grown at 490°C, Blue: Al seeded NW grown at 460°C followed by a rapid thermal anneal at 900°C, 10 s. (b) Al doping levels as a function of anneal temperature. The arrow shows the change in activation when a NW grown at 460°C is annealed to 900°C.

that the activated concentration shifts ~ six orders of magnitude following a 900°C anneal. This shows the utility of single NW UPS in determining industrially critical information on NW doping.

The smallest diameter structure measured with photoemission, a 1.7 nm single-walled carbon nanotube (SWCNT) was captured with this approach [41]. SWCNTs are actively under study for use in carbon electronics among other applications. Figure 22 shows a UPS spectrum from a 1.7 nm SWCNT comparing polarization parallel and perpendicular to the long axis of the tube. The binding energies of two observable van-Hove singularities, peaks in the one-dimensional density-of-states observed in the spectrum were determined.

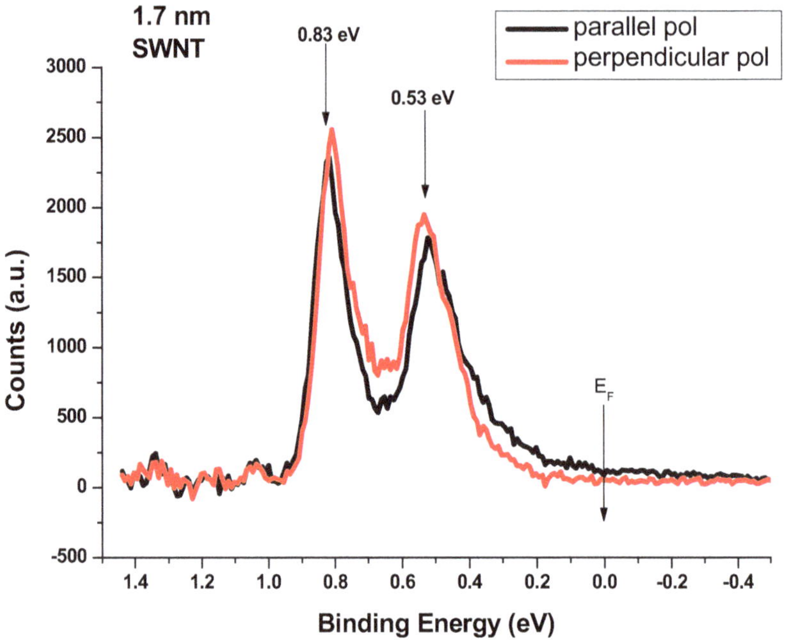

Fig. 22. Photoelectron spectra of an individual 1.7 nm SWCNT taken with light polarized parallel and perpendicular to the long axis of the CNT. The lack of polarization sensitivity is due to the strong effect of surface plasmons on the dielectric properties of the CNT.

References

[1] A. Einstein, "The photoelectric effect," *Ann. Phys.*, vol. 17, no. 132, p. 4, 1905.

[2] K. C. Pandey, J. L. Freeouf, and D. E. Eastman, "Photoemission and band-structure studies of the GaAs (110) surface," *J. Vac. Sci. Technol.*, vol. 14, no. 4, pp. 904–909, 1977.

[3] E. W. Plummer and W. Eberhardt, "Angle-resolved photoemission as a tool for the study of surfaces," *Adv. Chem. Physics*, vol. 49, pp. 533–656, 2007.

[4] D. Cahen and A. Kahn, "Electron energetics at surfaces and interfaces: Concepts and experiments," *Adv. Mater.*, vol. 15, no. 4, pp. 271–277, 2003.

[5] W. A. Harrison, *Solid State Theory*. Courier Corporation, Mineola, 1980.

[6] R. Haight, "Electron dynamics at surfaces," *Surf. Sci. Rep.*, vol. 21, no. 8, pp. 275–325, 1995.

[7] M. P. Seah and W. A. Dench, "Quantitative electron spectroscopy of surfaces: A standard data base for electron inelastic mean free paths in solids," *Surf. Interface Anal.*, vol. 1, no. 1, pp. 2–11, 1979.

[8] D. E. Aspnes and A. A. Studna, "Dielectric functions and optical parameters of Si, Ge, Gap, GaAs, GASB, INP, INAS, and INSB from 1.5 to 6.0 ev," *Phys. Rev. B*, vol. 27, no. 2, p. 985, 1983.

[9] R. G. Parr, "Density functional theory," in *Electron Distributions and the Chemical Bond*, P. Coppers and M. B. Hall (eds.), Springer, Berlin, 1982, pp. 95–100.

[10] I. M. Band, Y. I. Kharitonov, and M. B. Trzhaskovskaya, "Photoionization cross sections and photoelectron angular distributions for X-ray line energies in the range 0.132–4.509 keV targets: $1 \leq Z \leq 100$," *At. Data Nucl. Data Tables*, vol. 23, no. 5, pp. 443–505, 1979.

[11] K. Siegbahn, *ESCA; Atomic, Molecular and Solid State Structure Studied by Means of Electron Spectroscopy*, vol. 20. Almqvist & Wiksells, Stockholm, 1967.

[12] F. J. Himpsel, F. R. McFeely, A. Taleb-Ibrahimi, J. A. Yarmoff, and G. Hollinger, "Microscopic structure of the SiO_2/Si interface," *Phys. Rev. B*, vol. 38, no. 9, p. 6084, 1988.

[13] C. N. Berglund and W. E. Spicer, "Photoemission studies of copper and silver: Theory," *Phys. Rev.*, vol. 136, no. 4A, p. A1030, 1964.

[14] S. Tanuma, C. J. Powell, and D. R. Penn, "Calculations of electron inelastic mean free paths for 31 materials," *Surf. Interface Anal.*, vol. 11, no. 11, pp. 577–589, 1988.

[15] W. Eberhardt and F. J. Himpsel, "Dipole selection rules for optical transitions in the fcc and bcc lattices," *Phys. Rev. B*, vol. 21, no. 12, p. 5572, 1980.

[16] M. Lewenstein, P. Balcou, M. Y. Ivanov, A. L'huillier, and P. B. Corkum, "Theory of high-harmonic generation by low-frequency laser fields," *Phys. Rev. A*, vol. 49, no. 3, p. 2117, 1994.

[17] I. P. Christov, M. M. Murnane, and H. C. Kapteyn, "High-harmonic generation of attosecond pulses in the 'single-cycle' regime," *Phys. Rev. Lett.*, vol. 78, no. 7, p. 1251, 1997.

[18] W. Becker, A. Lohr, M. Kleber, and M. Lewenstein, "A unified theory of high-harmonic generation: Application to polarization properties of the harmonics," *Phys. Rev. A*, vol. 56, no. 1, p. 645, 1997.

[19] T.-C. Chiang, J. A. Knapp, D. E. Eastman, and M. Aono, "Angle-resolved photoemission and valence band dispersions for GaAs: Direct vs indirect models," *Solid State Commun.*, vol. 31, no. 12, pp. 917–920, 1979.

[20] R. Haight, J. A. Silberman, and M. I. Lilie, "Novel system for picosecond photoemission spectroscopy," *Rev. Sci. Instrum.*, vol. 59, no. 9, pp. 1941–1946, 1988.

[21] R. T. Williams, T. R. Royt, J. C. Rife, J. P. Long, and M. N. Kabler, "Picosecond time-resolved photoelectron spectroscopy of ZnTe," *J. Vac. Sci. Technol.*, vol. 21, no. 2, pp. 509–513, 1982.

[22] H. Petek and S. Ogawa, "Femtosecond time-resolved two-photon photoemission studies of electron dynamics in metals," *Prog. Surf. Sci.*, vol. 56, no. 4, pp. 239–310, 1997.

[23] W. S. Fann, R. Storz, H. W. K. Tom, and J. Bokor, "Direct measurement of nonequilibrium electron-energy distributions in subpicosecond laser-heated gold films," *Phys. Rev. Lett.*, vol. 68, no. 18, p. 2834, 1992.

[24] W. S. Fann, R. Storz, H. W. K. Tom, and J. Bokor, "Electron thermalization in gold," *Phys. Rev. B*, vol. 46, no. 20, p. 13592, 1992.

[25] D. Lim and R. Haight, "*In situ* photovoltage measurements using femtosecond pump-probe photoelectron spectroscopy and its application to metal–HfO2–Si structures," *J. Vac. Sci. Technol. A*, vol. 23, no. 6, pp. 1698–1705, 2005.

[26] C. G. Durfee III, A. R. Rundquist, S. Backus, C. Herne, M. M. Murnane, and H. C. Kapteyn, "Phase matching of high-order harmonics in hollow waveguides," *Phys. Rev. Lett.*, vol. 83, no. 11, p. 2187, 1999.

[27] R. Bartels, S. Backus, E. Zeek, L. Misoguti, G. Vdovin, I. P. Christov, M. M. Murnane, and H. C. Kapteyn, "Shaped-pulse optimization of coherent emission of high-harmonic soft X-rays," *Nature*, vol. 406, no. 6792, pp. 164–166, 2000.

[28] X. Zhang, A. R. Libertun, A. Paul, E. Gagnon, S. Backus, I. P. Christov, M. M. Murnane, H. C. Kapteyn, R. A. Bartels, and Y. Liu, "Highly coherent light at 13 nm generated by use of quasi-phase-matched high-harmonic generation," *Opt. Lett.*, vol. 29, no. 12, pp. 1357–1359, 2004.

[29] R. Haight and J. A. Silberman, "Surface intervalley scattering on GaAs (110): Direct observation with picosecond laser photoemission," *Phys. Rev. Lett.*, vol. 62, no. 7, p. 815, 1989.

[30] R. Haight, J. Bokor, J. Stark, R. Storz, R. R. Freeman, and P. H. Bucksbaum, "Picosecond time-resolved photoemission study of the InP (110) surface," *Phys. Rev. Lett.*, vol. 54, no. 12, p. 1302, 1985.

[31] S. B. Zhang and M. L. Cohen, "Surface states on GaAs (110)," *Surf. Sci.*, vol. 172, no. 3, pp. 754–762, 1986.

[32] J. L. Birman, M. Lax, and R. Loudon, "Intervalley-scattering selection rules in III-V semiconductors," *Phys. Rev.*, vol. 145, no. 2, p. 620, 1966.

[33] R. Haight and D. R. Peale, "Antibonding state on the Ge (111): As surface: Spectroscopy and dynamics," *Phys. Rev. Lett.*, vol. 70, no. 25, p. 3979, 1993.

[34] M. Probst and R. Haight, "Unoccupied molecular orbital states of tris (8-hydroxy quinoline) aluminum: Observation and dynamics," *Appl. Phys. Lett.*, vol. 71, no. 2, pp. 202–204, 1997.

[35] A. Rettenberger, P. Leiderer, M. Probst, and R. Haight, "Ultrafast electron transport in layered semiconductors studied with femtosecond-laser photoemission," *Phys. Rev. B*, vol. 56, no. 19, p. 12092, 1997.

[36] G. Rohde, A. Hendel, A. Stange, K. Hanff, L.-P. Oloff, L. X. Yang, K. Rossnagel, and M. Bauer, "Time-resolved ARPES with sub-15 fs temporal and near Fourier-limited spectral resolution," *Rev. Sci. Instrum.*, vol. 87, no. 10, p. 103102, 2016.

[37] A. L. Cavalieri, N. Müller, T. Uphues, V. S. Yakovlev, A. Baltuška, B. Horvath, B. Schmidt, L. Blümel, R. Holzwarth, and S. Hendel, "Attosecond spectroscopy in condensed matter," *Nature*, vol. 449, no. 7165, pp. 1029–1032, 2007.

[38] T. Rohwer, S. Hellmann, M. Wiesenmayer, C. Sohrt, A. Stange, B. Slomski, A. Carr, Y. Liu, L. M. Avila, and M. Kalläne, "Collapse of long-range charge order tracked by time-resolved photoemission at high momenta," *Nature*, vol. 471, no. 7339, pp. 490–493, 2011.

[39] S. Mathias, S. Eich, J. Urbancic, S. Michael, A. V. Carr, S. Emmerich, A. Stange, T. Popmintchev, T. Rohwer, and M. Wiesenmayer, "Self-amplified photo-induced gap quenching in a correlated electron material," *Nat. Commun.*, vol. 7, 2016.

[40] L. Miaja-Avila, C. Lei, M. Aeschlimann, J. L. Gland, M. M. Murnane, H. C. Kapteyn, and G. Saathoff, "Laser-assisted photoelectric effect from surfaces," *Phys. Rev. Lett.*, vol. 97, no. 11, p. 113604, 2006.

[41] S. Hellmann, T. Rohwer, M. Kalläne, K. Hanff, C. Sohrt, A. Stange, A. Carr, M. M. Murnane, H. C. Kapteyn, and L. Kipp, "Time-domain classification of charge-density-wave insulators," *Nat. Commun.*, vol. 3, p. 1069, 2012.

[42] R. Haight, G. Sirinakis, and M. Reuter, "Photoelectron spectroscopy of individual nanowires of Si and Ge," *Appl. Phys. Lett.*, vol. 91, no. 23, p. 3116, 2007.

[43] R. S. Wagner and W. C. Ellis, "Vapor–liquid–solid mechanism of single crystal growth," *Appl. Phys. Lett.*, vol. 4, no. 5, pp. 89–90, 1964.

[44] J. Westwater, D. P. Gosain, S. Tomiya, S. Usui, and H. Ruda, "Growth of silicon nanowires via gold/silane vapor–liquid–solid reaction," *J. Vac. Sci. Technol. B Microelectron. Nanom. Struct. Process. Meas. Phenom.*, vol. 15, no. 3, pp. 554–557, 1997.

[45] B. A. Wacaser, M. C. Reuter, M. M. Khayyat, C.-Y. Wen, R. Haight, S. Guha, and F. M. Ross, "Growth system, structure, and doping of aluminum-seeded epitaxial silicon nanowires," *Nano Lett.*, vol. 9, no. 9, pp. 3296–3301, 2009.
[46] Carlos Aguilar, Richard Haight, "Single Nanowire Photoelectron Spectroscopy", in *Handbook of Instrumentation and Techniques for Semiconductor Nanostructure Characterization*, R. Haight, F. M. Ross, and J. B. Hannon (eds.) World Scientific Press, 2011.

Chapter 5

Femtosecond Ablation and Mask Repair

5.1 Introduction

Lasers have historically been used as devices for melting and cutting [1, 2]. Because of the ultra-narrow bandwidth of light emitted from a laser oscillator, as well as the directed nature of the light emission from the laser cavity, focusing to spot sizes given essentially by the diffraction limit of the optics is possible. The diffraction limited spot size is given by $\sim \lambda/(2n \sin \Theta)$, where λ is the wavelength of the light being focused, n is the refractive index of the medium in the region of the focus, and Θ is the maximal angle of the focused cone of light. The term $n \sin \Theta$ is called the numerical aperture (NA) and is a standard parameter in optics. In air the NA can approach 1, but can be $\sim$1.5 when light is focused in a liquid environment that utilizes high index fluorinated oils. As an example, a high quality, aberration corrected microscope objective with a NA = 0.95 can resolve features down to $\sim\lambda/2$.

An advantage of lasers is that the light produced is typically spectrally pure (extremely narrow bandwidth), dramatically reducing chromatic aberration. Chromatic aberration is one of the many factors that degrade the performance of an optical system. It derives from the fact that the index of refraction of a transmissive optical element, such as a lens, will change with wavelength. For a given bandwidth of light, each wavelength comprising the light will be focused at a slightly different distance from the lens thereby degrading both the radius and depth of the focal spot. In a multi-elemental microscope objective, multiple lenses with different properties, including certain elements comprising different types of glasses, are utilized to compensate for chromatic and other aberrations. Spherical aberration and astigmatism are also lens issues that can degrade focal performance.

The process of ablation starts with the absorption of intense pulses of light. Consider, for example, an intense pulse of laser light impinging on a solid material [3]. Absorption of visible photons in the light pulse by electrons in the material raises those electrons to excited states of the material. If the pulse is weak, those electrons will relax back to their ground state through the emission of phonons — vibrational modes of the atoms in the material; the main consequence of this is to heat the material. The amount of actual heating of course depends on the laser intensity. Recalling our discussion in Section 4.8 involving laser excitation and subsequent relaxation of electrons in a Au film, the fundamental timescale for electron–phonon scattering in a metal is several picoseconds [4], while for a semiconductor it is a few hundred femtoseconds. It is similarly short in insulators. For all of these materials under infrared, visible or ultraviolet (UV) light illumination, it may take many electron–phonon interactions to relax the electronic system to its ground state thereby producing enough phonons to heat the solid. In semiconductors, the final ground state is reached by the emission of photons as electrons and holes recombine across the band gap.

Now consider a far more intense pulse of laser light such that the high number of photoexcited electrons produces an even higher number of phonons. For a sufficiently intense pulse of light the temperature may rise many hundreds of degrees, enough to melt the material. If the pulsewidth of the light is relatively short, but not ultrashort, say hundreds of picoseconds to nanoseconds or more, the material will simply melt. Laser melting has been studied in great depth in the literature [2]. Melting may be the desired goal in a laser–solid interaction. An example of this, used widely in semiconductor manufacturing, is laser annealing of a semiconductor implanted with a dopant where it is highly desirable to achieve "activation" without significant diffusion of the dopant atom outside a specific spatial region. Here, activation refers to the replacement of the host atom with the dopant atom resulting in donation (n-type), or acceptance (p-type) of an electron within the host semiconductor.

Beyond melting there is removal of material, commonly referred to as ablation with applications in high spatial resolution machining. Using longer laser pulses can certainly ablate materials but there is attendant collateral "damage". By damage here we mean that there is modification of the material in the region beyond the intended focal area of the laser light. The damage may take the form of melted material that is ejected, or splattered about. Damage can also occur from shockwave propagation [5] in the regions near to the focus of the laser since the heating, melting, and vaporization of the material is a violent process.

Damage from shockwave propagation and splatter may not be an issue for low-resolution material machining, but in applications where this is undesirable we can exploit the fundamentally different physics associated with femtosecond pulsed laser ablation to achieve significantly better machining

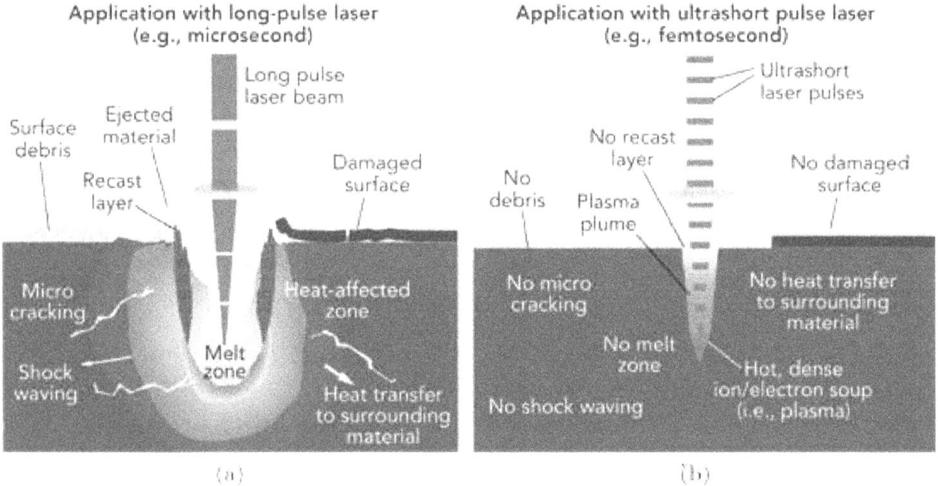

Fig. 1. Schematic comparing long pulse laser ablation (a) and ablation with femtosecond lasers (b). (Reproduced with permission from G. Shannon).

results. Initial femtosecond studies were carried out on metal and silica surfaces [6–9]. Some direct comparisons can be made by looking at Fig. 1. Figure 1(a) is a schematic of a long pulse laser, e.g. nanoseconds or longer. For a sufficiently intense pulse, material in the focal region will be melted and then vaporized. The material evolves through sequential phases that include heating, melting, and vaporization with the production of ejected material. As shown schematically, there is a localized melt zone around the focal region that results from diffusion of thermal energy from the focal region. There is also a "heat-affected zone" where high temperatures can modify the properties of the material. Surface debris appears in the form of splattered material and the propagation of a shock wave can produce microcracking of the material.

In comparison, as shown in Fig. 1b when material is ablated with a femtosecond pulse of light the rapid introduction of energy essentially converts the material to a plasma before significant thermal diffusion can occur. As a result, there is little or no shockwave production, melting, splatter, or surface damage. As we will describe in greater detail in this chapter, this process when applied to the removal of absorptive defects from a photomask used in the printing of integrated circuitry, produces nearly error-free, debris-free repairs with no damage to the underlying glass.

5.2 Some Basics of Femtosecond Laser Ablation

How quickly can we introduce energy into a system relative to fundamental response times of the material? If we consider some timescales relevant to our discussion, the physics of femtosecond laser ablation comes more into focus. The key issues are the duration of the light pulse relative to specific material time constants. When a pulse of light is absorbed it is initially the electronic system that is excited. Excited electrons scatter amongst themselves thermalizing their energy on extremely short timescales, typically femtoseconds. For example, the electron plasma frequency, dependent upon the density of electrons is given by [3]

$$\omega_{\mathrm{pe}}^2 = n_e e^2/m_e \varepsilon_{\mathrm{o}}, \tag{1}$$

where n_e is the electron density, m_e is the electron mass and ε_{o} is the free space permittivity. $\omega_{\mathrm{pe}}^{-1} \ll 1\,\mathrm{fs}$ for the intensities required for ablation and this is much shorter than the typical femtosecond laser ablation pulse.

In a dense plasma created during a femtosecond laser pulse [10, 11], the electron–phonon collision frequency is given by

$$\omega_{e-\mathrm{ph}} \approx [m_e/M_{\mathrm{ion}}]^{1/2} * \left(\frac{\mathrm{IP}_{\mathrm{solid}}}{\hbar}\frac{T}{T_D}\right), \tag{2}$$

where $\mathrm{IP}_{\mathrm{solid}}$ is the ionization potential of the material, T_D is the Debye temperature and m_e and M_{ion} are the electron and ion masses, respectively.

The electron–ion energy transfer time in a dense laser produced plasma is given by

$$\tau_{e-i} \approx \frac{M}{m_e} \Bigg/ \left[2 \times 10^{-6} \frac{Z n_e}{T_{eV}^{\frac{3}{2}}}\right], \tag{3}$$

for materials ranging from metals through semiconductors and even dielectrics $\tau_{e-i} \sim$ 3–6 ps. This indicates that for 100 fs laser pulses that are routinely produced in commercially available amplified femtosecond lasers, the ions are in a relatively low energy state, i.e. they are not hot.

So then, what actually drives the ions out of the sample in the ablation process? Since the high-intensity, short-pulsed laser drives electrons both into antibonding states of the material as well as into the vacuum, those electrons are no longer available to bond the ion cores and extremely strong electrostatic forces drive the ions out of the material. This is the reason that conventional thermal melting and diffusion of heat into surrounding areas of the material is avoided with femtosecond ablation and results in an ablated volume that closely matches the focal region of the laser. In cases where only a small fraction of the laser pulse is above the threshold for ablation, sub-diffraction limited material removal [12] is possible, as we will discuss in this chapter.

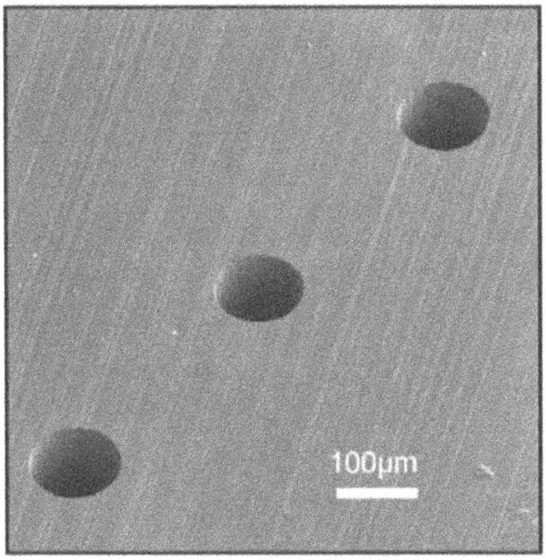

Fig. 2. Holes in stainless steel created with 150 fs, 780 nm laser pulses. (Kamlage, Bauer, Ostendorf, Chickov, App. Phys. A, 77, 307 (2003) DOI:10.1007/s0339-003-2120-x).

The obvious use of an intense femtosecond focused laser pulse is to drill holes in materials, and indeed this is often the application. A typical example of a hole created with femtosecond laser ablation is shown in Fig. 2.

But a perhaps more subtle and surgical application involves removing one material from another. In later sections, we will discuss femtosecond ablative repair of photomasks. This particular case is an extremely good example of the advantages of femtosecond pulses for machining, in that it combines extraordinarily high spatial resolution and limited, almost non-existent collateral damage associated with longer laser pulses. But first lets discuss the optics involved in focusing a laser pulse to a small spot.

5.3 Gaussian Optics and the Focusing of a Laser to a Small Spot

To understand the nature of focusing of a laser beam by a high quality, aberration–corrected microscope objective, particularly for the purpose of ablating material, we present a short discussion of Gaussian optics and the propagation of a Gaussian laser beam.

There are several initial considerations when discussing Gaussian beams. The typical output of a continuous-wave (CW) laser is highly coherent and the emergent wavefront can be precisely defined. The output of a transverse electromagnetic mode (TEM_{00}) laser beam can be described as having a Gaussian intensity profile perpendicular to its direction of propagation as the beam of light emerges from a laser cavity or focal region [13]. This is our starting point. The diameter is defined as twice the radial distance from the center of the beam where the transverse field amplitude drops to $1/e$ of its peak value. The electric field of the laser light is given by

$$E = E_0 e^{-\left(\frac{r^2}{w^2}\right)} \tag{4}$$

and the intensity profile is therefore

$$I = I_0 e^{-\left(\frac{2r^2}{w^2}\right)} = \frac{2P}{\pi w^2} e^{-\left(\frac{2r^2}{w^2}\right)}, \tag{5}$$

where r is radial distance from the beam axis and w is the distance from the beam axis where the irradiance drops to e^{-2} of its value. P is the total power in the beam. A cross-sectional profile of a Gaussian beam is shown in Fig. 3 along with a schematic picture of the wavefronts as the beam transits to and then emerges from its focus.

Consider in Fig. 4 where *w(z)* depends on the distance z that the beam has propagated. At the waist $(2w_0)$ of the beam the wavefront is flat but as the beam propagates from the waist the wavefront becomes spherical. The value of w_0 here can be the radius of the laser beam at the output of the laser or the radius of the beam at the focus of a lens. The mathematical form for the magnitude of the waist as a function of distance from the point where the wavefront is flat is given by

$$w(z) = w_0 \left[1 + \left(\frac{\lambda z}{\pi w_0^2}\right)^2\right]^{1/2}, \tag{6}$$

where for large distances

$$w(z) \cong \frac{\lambda z}{\pi w_0}. \tag{7}$$

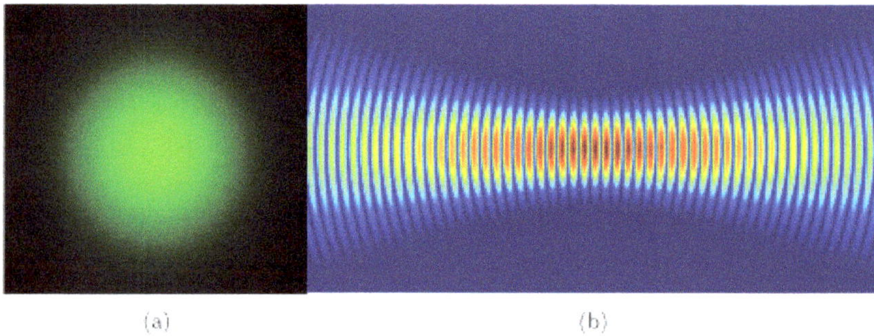

Fig. 3. (a) radial intensity profile of a Gaussian beam. (b) wavefront curvature in the region at and near the focus.

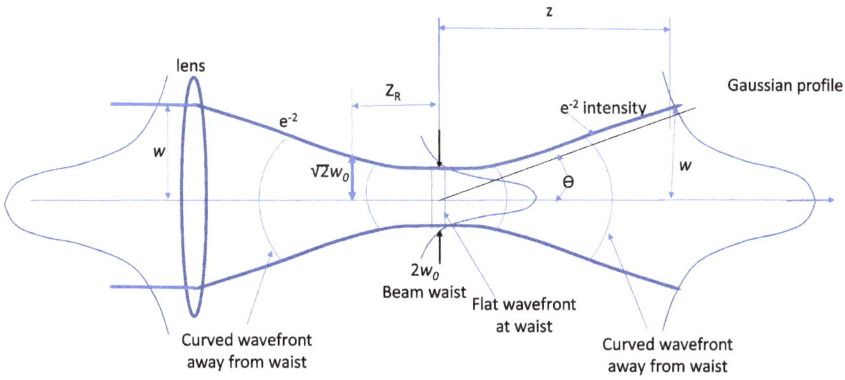

Fig. 4. Schematic showing the focusing of a laser beam with a Gaussian radial profile. Input beam radius w is incident on a lens that focuses the beam to a waist whose diameter is 2 w_0.

For Gaussian beams, near the waist the divergence angle is very small (as seen in Fig. 4) but well away from the w_0, the angular divergence of the light cone approaches an asymptotic value of Θ as shown in Fig. 4 (solid black line) and this value is given by

$$\theta = \frac{w(z)}{z} = \frac{\lambda}{\pi w_0}. \tag{8}$$

A convenient parameter that characterizes the distance over which the beam radius spreads by a factor of $\sqrt{2}$ is given by

$$Z_R = \frac{\pi w_0^2}{\lambda}, \tag{9}$$

where Z_R is the Rayleigh range, so that $w(z)$ in terms of Rayleigh range is

$$w(z) = w_0 \left[1 + \left(\frac{z}{Z_R}\right)^2\right]^{1/2}. \tag{10}$$

For our interests here, we are concerned with focusing of a Gaussian beam to a small spot, concentrating the beam to achieve the highest spatial resolution possible while ablating the intended material. The standard lens equation is written as

$$\frac{1}{o} + \frac{1}{i} = \frac{1}{f}, \tag{11}$$

where o is the object distance, i is the image distance, and f is the focal length of the lens being employed. For Gaussian beams this simple equation takes a different form, derived by Self [13] as

$$\frac{1}{i + \frac{z_R^2}{i-f}} + \frac{1}{o} = \frac{1}{f}. \tag{12}$$

To focus a Gaussian beam in order to concentrate the light for high resolution ablation, consider a collimated beam illuminating a lens with waist w as shown in Fig. 4. The angle with which the light is concentrated is Θ and since for small angles $\tan\Theta \sim \Theta$ we find that the waist at the focus is

$$w_0 = \frac{\lambda f}{\pi w}, \tag{13}$$

where w_0 is the waist at the focus and w is the radius of the beam illuminating the lens. As an example to be discussed in greater detail when we describe photomask repair, a microscope objective [14] with an input aperture of 3 mm and a focal length of 1.33 mm focuses a 250 nm laser beam to a diameter of 140 nm. As we will show, this is the optical spot size, but the nonlinear nature of material ablation can be exploited to ablate even smaller features through careful control of the laser pulse intensity.

The depth of focus (dof) is another important parameter. For focusing of a microscope image with an objective lens this essentially corresponds to the range of focal positions of the object relative to the imaging objective where the image remains in focus. This is a bit arbitrary, and different criteria are often used. If the dof is defined as the distance between the two values of the beam waist where the beam is $\sqrt{2}$ larger than at w_0 then for a beam width w incident on a lens of focal length f

$$\text{dof} = \frac{2\lambda f^2}{\pi w^2}. \tag{14}$$

This value is also known as the confocal parameter or "b" value. Using our previous example of focusing through a deep UV (DUV), high NA objective we find that the dof is ~160 nm. This is quite a small range and requires a high degree of mechanical control of the focusing system to achieve. As a result high quality microscopes (e.g. Zeiss, Leica, Nikon) with exceptional focusing capabilities are required for such focal precision.

5.4 Photomasks and Femtosecond Photomask Repair

A prime application of femtosecond light pulses in the industrial setting is the "machining" of materials. Machining can take the form of hole drilling, sculpting of patterned materials or patterning of the material itself. As we have stated above, for pulses substantially longer than characteristic electron–phonon scattering times (~1 ps for metals and hundreds of femtosecond for semiconductors), laser pulses will drive the absorber through heating, melting, and vapor phases. This typically gives rise to a combination of deleterious effects such as metal splatter, shock wave propagation and substantial heat diffusion radial to the focal region. Metal splatter, as we will describe below in applications to mask repair can coat otherwise pristine glass with metal droplets.

It should be mentioned that when we reference "glass" for photomasks, we actually mean fused silica. The crystalline form of fused silica is quartz or purely SiO_2. The more generic term for glass refers to SiO_2 with various contaminants. For example, a common form of glass is soda–lime glass that possesses Na, Ca, Ti, Mg, Al, and Fe. Such contaminants introduce a cutoff in transmission below ~350 nm; as a result, these glasses are incompatible with UV and DUV lithography. Given that modern photolithography utilizes 193 nm light from an ArF excimer laser, only fused silica substrates can be utilized as they are transparent at this wavelength.

Returning to our discussion of ablation, shockwave propagation derived from long pulse lasers can damage adjacent structures. Heat diffusion can drive absorber atoms into regions where they should not be. Probably the key advantage of using femtosecond pulses is their ability to drive the absorbing material in the focal region directly into the plasma state, bypassing a melt phase with minimal heating of adjacent material. This occurs because a high-density of valence electrons are excited simultaneously, in times well below the electron–phonon scattering time. As a result, the material disassembles directly, before the excess electron energy can be transferred to the material phonon system as heat. This is a key reason that femtosecond pulses ablate so accurately and, as we will show, not only achieves classical diffraction limited resolution, but can ablate material at even smaller dimensions under appropriate conditions.

The process of printing circuit patterns on chips is analogous to the more familiar commercial photographic process. Although current photography is primarily digital, with images captured by CCD arrays, in days of yore actual photographic film was used. The process of snapping a picture with a standard camera involved opening a shutter in the camera for a set duration; the lens system of the camera formed an image on film at the back of the camera. This image is actually the negative and within a tool or optical lab, light would be projected through the negative onto appropriately coated paper to create the print.

In analogy with the photographic steps described above, the process for printing chips is quite similar and is described briefly here. In essence, the photographic negative is replaced by a patterned photomask and the "print" is a Si wafer coated with a photoresist, typically a photosensitive polymer film. The pattern on the photomask represents one level of circuitry to be transferred to the Si wafer; in modern chip manufacturing there may be as many as 50 or more levels of circuitry to produce a functioning microprocessor. There are many types of photomasks used for lithographic manufacture of chips but the simplest and mostly widely used are "chrome-on-glass" shadow masks. Essentially, a 6″ ($152.4\,\mathrm{mm}^2$) square of exceptionally high quality fused silica glass is coated with a chrome alloy; this is called a blank. The details of the alloy itself are usually closely held secrets of the blank manufacturer but typically the absorber film is a chrome-oxynitride. The blank is coated with either an optical or electron sensitive polymer resist. The actual circuitry for a particular chip level is written by exposing the mask to light or electron beams, depending on the spatial resolution required (Fig. 5). In the case of electron beam writing, an "e-beam" is scanned over the mask with extraordinary precision. The pattern is created in the resist by modulation of the e-beam current focused to nanometer scales as it scans over the mask. Mask writing tools are very complex and expensive, costing as much as \$10–20 million. The writing process can take many hours, consuming sometimes as much as an entire day to transfer the circuit pattern to a single mask. Following e-beam or optical writing, the pattern must be developed. All of these mask manufacturing steps impart considerable value to the written mask and when inevitable errors during the write or develop

Fig. 5. Typical chrome-on-glass photomask.

steps of the mask manufacture occur, it is far more cost-effective to repair the defect than it is to rewrite the mask.

As mentioned, once the mask has been exposed to the patterning e-beam or light, the resist is "developed". For the case of a positive resist, the regions of the polymer exposed to the writing beam become soluble in the developer solution, while a negative resist becomes insoluble in the photoresist developer. Once the patterned resist is removed in the developer the exposed chrome is removed via a wet chemical etch process or by reactive ion etching (dry etch) to expose the fused silica substrate below. Finally, the rest of the polymer resist is removed and the resulting circuitry pattern is revealed. The mask is then cleaned and inserted into an optical exposure tool called a "stepper" [15]. The stepper is a tool that combines a multi-elemental optical objective, typically of high NA optimized for a single wavelength, with a nanometer scale precision mechanical positioner. Typical stepper optics that operate in air have NAs just below 1, while optics that work in high index transparent fluids, called immersion optics [16], possess NAs as high as 1.5. Both approaches are designed to achieve the highest possible spatial resolution. The positioning stage locates the exposure field under the objective lens. To achieve the highest resolution images at the Si wafer surface, line-narrowed KrF (248 nm) or ArF(193 nm) excimer lasers are used. Shorter wavelength excimers exist such as the F_2 laser that operates at 157 nm [17] but have not been successfully utilized in industry. Extreme UV (EUV) lithography [18] is another approach that operates at 13.5 nm and is an entirely different approach that will be discussed later.

The lenses in the objective of the stepper have been precisely manufactured and coated with antireflective films to achieve as close to the classical diffraction limit as possible. The resolution limit, given by the ability of the lithographic system to print two distinct features in a resist coated wafer, is given approximately by

$$R = K * \lambda/\mathrm{NA}. \tag{15}$$

Here, NA is the numerical aperture of the system given by: $\mathrm{NA} = n * \sin\Theta$, where n is the index of refraction of the medium between the lens and the Si wafer and Θ is the half angle of the lens focus as described previously. When an immersion fluid is used, both the objective and the Si wafer are immersed in the dielectric fluid. The refractive index of these fluids at 193 nm (ArF excimer) can be in excess of 1.4. In addition, note that we have inserted a variable which is called the K-factor. K is a process-related parameter that includes various resolution enhancements such as multipole illumination, the use of chemically amplified resists, the employment of immersion optics, and more to achieve the absolute highest spatial resolution possible [19]. For 193 nm light and a K value ∼1 we can see that $R \sim 100$ nm for an air-based system while for an immersion system $R \sim 65$ nm and with an additional K enhancement where K can range as low as 0.25, $R \sim 35$ nm is possible. To achieve even higher spatial resolution additional methods such as the use of phase-shift masks, off-axis illumination, and multiple exposure modalities must be used. Finally, in order to achieve even higher spatial resolutions ranging to and below 10–15 nm, EUV sources and steppers must be used. These systems are just now beginning to reach production capability although throughput is low relative to conventional optical steppers.

5.5 Femtosecond Photomask Repair

As stated above, considering the expensive fabrication of a photomask, some of which may cost upward of $100 K to produce, the inevitable appearance of defects must be dealt with. These defects typically take the form of excess absorber in regions where there should only be clear glass. Such defects, when transferred to a Si wafer can result in electrical shorts, open or narrowed lines that can inhibit or completely prohibit current flow and render the chip inoperable.

The simplest method to deal with defects is to discard the mask and start over but this is prohibitively expensive particularly when as many as 30–50 or more photomasks are required to complete all of the circuitry levels in an advanced chip. Other repair techniques have been developed. These include nanosecond laser ablative removal of an absorber defect, focused ion beam etching and e-beam stimulated etching. Each of these approaches has significant problems. Nanosecond ablation melts the metal before driving it to evaporate and it is this melted stage that is problematic. Melted metal can evaporate explosively spreading molten droplets around the mask (see Fig. 6). In addition, the melted metal can diffuse into the glass permanently decreasing its transmission. For earlier nodes with spatial features larger than 200 nm this process was acceptable but became increasingly unacceptable as the repaired regions and splattered metal droplets were of the same size scale as the actual circuit patterns. Several examples and comparisons of nanosecond versus femtosecond ablation are shown below.

In Fig. 6, we display scanning electron micrographs (SEM) of the results from nanosecond laser ablation in Fig. 6(a) and femtosecond ablation in Fig. 6(b). These SEMs are collected from chrome-on-glass test photomasks. In Fig. 6(a), one vertical chrome bridge has been ablatively removed in the right region of figure with nanosecond laser pulses. As can be seen there is considerable metal splatter and curling of the edges of the horizontal chrome

(a)

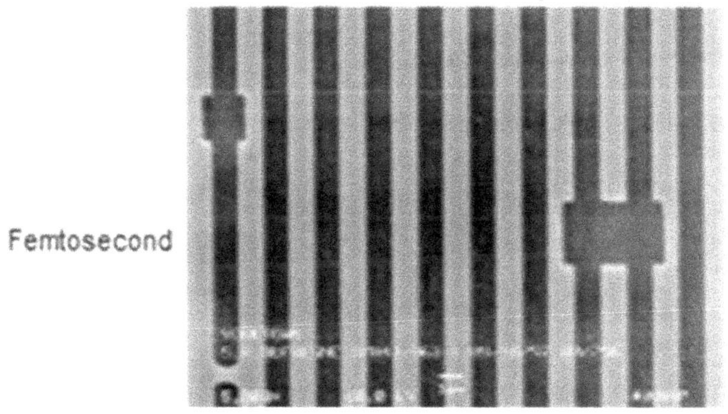

(b)

Fig. 6. SEM micrographs showing the comparison of ablative removal of a vertical chrome bridge line (a) using nanosecond laser pulses and (b) using femtosecond laser pulses. Note that the nanosecond laser pulses remove the bridge but result in considerable metal splatter and curling of the horizontal chrome lines.

lines. In Fig. 6(b) a similar mask has been ablated with femtosecond laser pulses. Note the surgical removal of chrome without the attendant splatter of metal.

Figure 7 shows the impact of nanosecond versus femtosecond pulses on the optical quality of the glass after chrome has been removed. The image was taken with an optical microscope in transmission with 248 nm light. Bright areas are regions on the mask devoid of Cr absorber. Figure 7(a) shows ablations using nanosecond laser pulses. Obvious damage to the glass has occurred in the form of "staining" that occurs due to diffusion of heated metal atoms into the underlying glass. This effect is essentially irreparable. Furthermore, some of the glass is also eroded giving rise to the subtle diffraction lines observable in the optical micrograph. This process, known

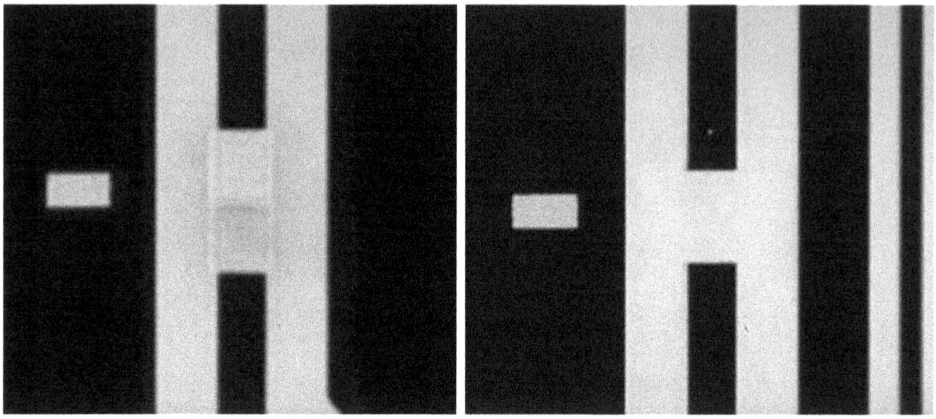

(a) (b)

Fig. 7. (a) Transmitted light optical micrograph of a chrome-on-glass photomask ablated with nanosecond laser pulses. Darkened region and interference fringes from etched fused silica can be seen. (b) Femtosecond-laser ablated region showing none of the interference fringes or loss of transmission.

as "riverbedding" will inevitably print unwanted patterns on the Si wafer, similar to the diffractive pattern observed in Fig. 7.

Focused ion beams also suffer from similar problems. While splattered metal droplets are typically not an issue, focused ion beams result in the implantation of Ga (the ions used to etch the defect). This leads to decreased transmission in the repaired region, similar to nanosecond repair. Even after methods to limit this problem were introduced, the attendant issue of riverbedding, where the underlying glass was sputtered as well became a significant problem. This form of glass erosion, similar to that observed with nanosecond laser ablation leads to phase shifts and interference between light that impinges on the unrepaired and repaired glass owing to the difference in glass thickness.

Figure 7(b) shows the same photomask after ablation with femtosecond pulses. In addition to clean removal of the Cr absorber, none of the glass damage in the form of in-diffused Cr or glass erosion is visible.

5.6 An Industrial Femtosecond Photomask Repair Tool

As described in the beginning of this chapter, when electrons in a material absorb light they are promoted to excited states of the system. Relaxation mechanisms such as phonon scattering convert excess electronic energy into lattice heat. In semiconductors, electrons that are excited into conduction states of the system lose their excess energy ($E - E_{\text{cbm}}$, where E is the initial electron energy in the conduction band and E_{cbm} is the conduction band minimum) through the emission of phonons, producing heat. In metals, phonon emission is also responsible for energy relaxation after initial excitation. Femtosecond photoelectron spectroscopy experiments helped to establish the picosecond cooling time in metals [4]. In semiconductors, electron–phonon scattering times are on the order of several hundred femtoseconds [20–22]. In ablation with pulses substantially longer than the characteristic electron–phonon emission time, the excess electron energy is converted to heat which melts and then evaporates the material.

As mentioned above, there exist a number of experiments carried out in semiconductors and metals that indicate that the electron–phonon scattering times for a wide range of materials is at and below 1 ps [23]. As a result, we can conclude that minimizing thermal processes such as thermal diffusion, in the ablative removal of material requires that energy is deposited in a time substantially below the electron–phonon coupling time. Using femtosecond pulses, we observe that the edges of the ablated regions are sharp, and there is no evidence of metal splatter or debris. The absorption of the ultrashort laser pulse creates a plasma whose ejection carries off a substantial fraction of the deposited laser energy, thereby minimizing residual thermal effects in the substrate.

The advantages of such an ultrafast ablative process are numerous. In addition to producing a sharp boundary between ablated and remaining material, the lack of substantial heat transfer to the substrate results in high optical transparency in the underlying fused silica. As mentioned above, it is frequently observed in nanosecond repair that metal diffuses into the underlying glass resulting in reduced transparency and pitting due to removal of the now-absorbing glass substrate. Conversely, no diffusion of metal is observed with femtosecond pulses and glass ablation has been measured to be $<$3–5 nm in depth. Additionally, since thermal diffusion is virtually eliminated, repaired regions could be scanned multiple times to improve the spatial and optical qualities of the repair without concern for cumulative damage to the underlying glass.

Utilizing the concepts and results of our discussions above, the gains realized with femtosecond pulses imply that highly accurate repairs could be achieved. To exploit these advantageous properties, a tool was constructed and installed in manufacturing at IBM's mask manufacturing plant in Burlington, Vt. A schematic of the manufacturing tool is displayed in Fig. 8.

There are several subsystems and we describe them here [12, 24, 25]. A highly accurate granite air-bearing stage, with $\pm$4 nm position stability

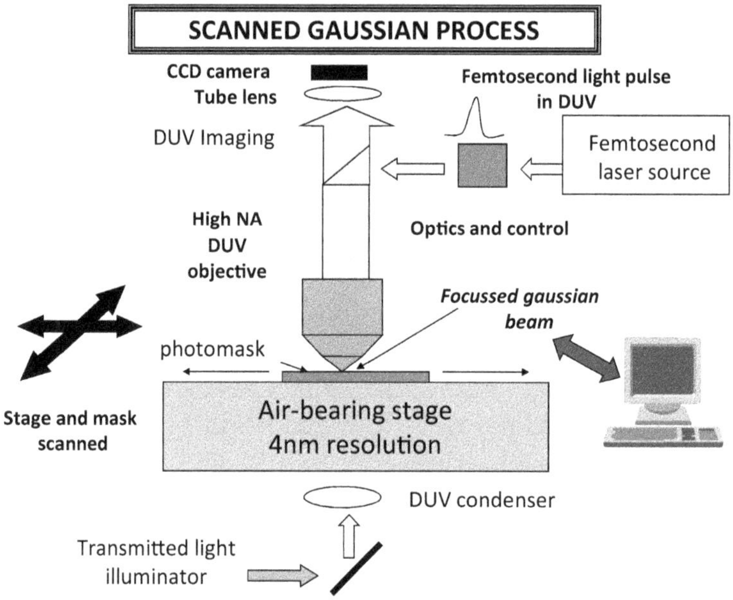

Fig. 8. Schematic of the femtosecond photomask repair tool — MARS.

positions the photomask under a 0.9 NA (Leica) DUV microscope objective [14]. The mask is illuminated in transmission with 248 nm light and the image of the mask is projected onto a CCD camera via a tube lens that resides above the objective. A 1 kHz repetition rate, regeneratively amplified Ti:sapphire laser that produces 120 fs pulses at 800 nm is frequency tripled to produce pulses at 266 nm, well within the transmission bandwidth of the microscope objective. The light is spatially filtered through a diamond pinhole to provide a uniform Gaussian mode and is coupled into the objective and projected onto the photomask surface. The beam is electro-optically shuttered and both stage and laser are controlled by a computer. In the design of the Burlington tool, called Mask Advanced Repair System (MARS), a high resolution microscope manufactured by Leica was installed. Modifications of the optical path and tube lens were made in order to incorporate a beam splitter that allowed viewing of the mask and the femtosecond laser beam. In daily operation, a mask repair technician delineates the region to be repaired on a computer display, and the mask is rastered under the irradiating laser pulses, which are focused to a Gaussian spot on the mask, until the defect is removed. Typically, <150 nm lines and holes are machined into the Cr absorber patterned on a standard Cr-on-glass photomask representing the standard ablation resolution achieved on a daily basis by Mask House technicians.

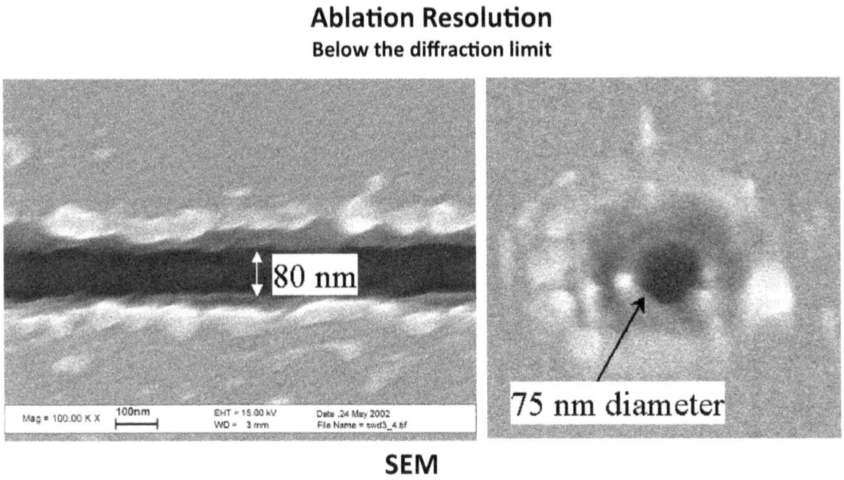

Fig. 9. SEMs of an 80 nm line and 75 nm hole in chrome-on-glass photomask ablated with near threshold intensity femtosecond 248 nm light pulses.

Since the ablation process is highly nonlinear, it is possible to achieve substantially better spatial resolution by adjusting the focal intensity to a value near the threshold for ablative removal of the absorber. Recalling our previous discussion on the Gaussian focal spot size (Eq. 13) at a wavelength of 250 nm using a 0.9 NA DUV microscope objective we calculated a diffraction limited focal diameter equal to 140 nm. Sub-diffraction limited ablation can be achieved by careful control of the pulse intensity and this resolution is displayed in Fig. 9, where an SEM of an ablated line whose width is ~80 nm is displayed. In the adjacent SEM, a 75 nm hole diameter was achieved, corresponding to $\lambda/3$ and substantially below the diffraction limit.

Figure 10 displays an example of the extreme accuracy with which line edges can be reconstructed. Figure 10(a) shows the initial transmitted line of 1,200 nm width. In this experiment, the focused Gaussian beam was used to ablate a series of iterative cuts or "nibbles" in Cr [12, 25] along the edge of an existing feature on the photomask; the width of the line was measured after each ablative sweep. The stage was then moved by one pixel on the computer monitor corresponding to 22 nm on the photomask. A series of 22 nm steps, ablations and measurements were carried out and are displayed in Fig. 10(b) where the linedwidth as a function of the cumulative number of trims is plotted. The data are well fit by a line whose slope is 22.7 nm very close to the intended 22 nm step size. Importantly, the RMS edge error over 35 trims is only 5.1 nm. Micron and sub-micron size defects can be removed typically in seconds to minutes. High throughput coupled with precise spatial control

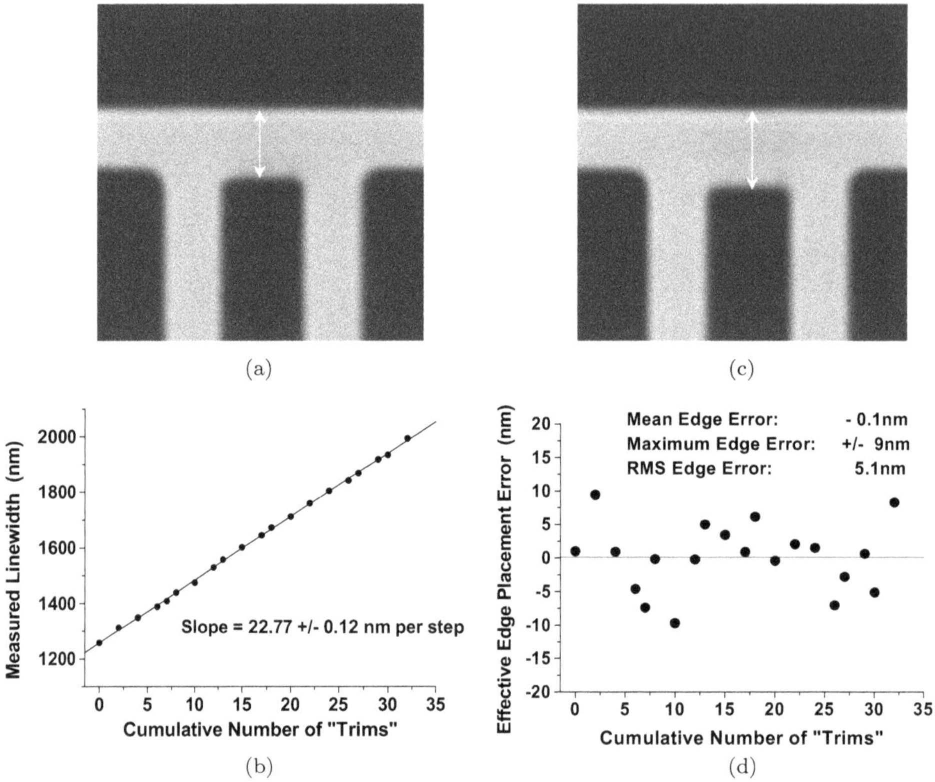

Fig. 10. (a) Transmitted 248 nm optical micrograph of a chrome-on-glass photomask. The top line is 1,200 nm wide. (b) measured linewidth versus trims. The intended increase in each trim was 22 nm. (c) final linewidth achieved. (d) edge placement error versus trims. Over 35 trims the RMS edge error was 5.1 nm.

and minimal damage to the underlying fused silica substrate are all critical requirements that femtosecond repair has successfully met in manufacturing on a daily basis.

Further evidence of the efficacy of employing femtosecond pulses for ablative mask repair is shown in Fig. 11. This figure displays the Aerial Imaging Measurement System (AIMS) [26] measurement, utilizing 193 nm light, of a vertical defect in the third horizontal line (counted from the top) of a feature on a photomask. AIMS measurements involve collecting images through a microscope as the objective focus is shifted from above to below the actual in-focus condition for the feature on the mask. In this manner phase information on the mask, defects, and subsequent repair can be evaluated. The AIMS measurement tool not only determines overall transmission at focus but if erosion of the substrate glass occurs in a repair, diffraction patterns associated with different glass heights in and around the repair

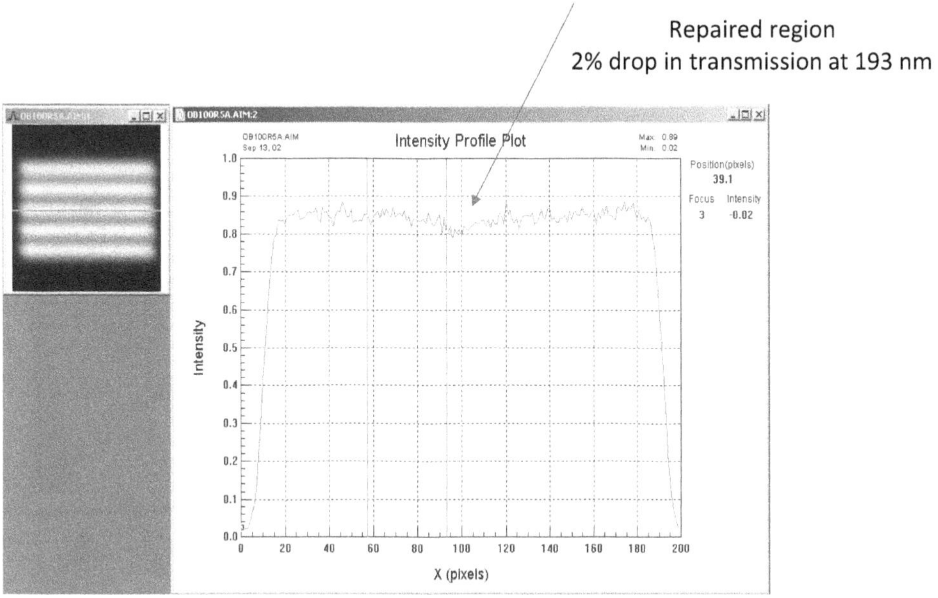

Fig. 11. AIMS measurement of the repaired area of a "bridge" defect (along the third line from the top). Repaired area showed only a 2% loss in transmission at 193 nm.

will be observed. AIMS measurements are a standard throughout the mask industry.

AIMS operating at 248 nm was unable to detect the defect repaired with 266 nm femtosecond pulses, indicating that the transmission through the repaired area at 248 nm was at or near 100%. For the mask to be utilized in a 193 nm stepper, an AIMS measurement was required at 193 nm and that is shown in Fig. 11. The repaired defect can be just seen in the 193 nm AIMS image. The repair of the feature was carried out with femtosecond pulses at 266 nm. Using the 193 nm AIMS tool we note a small depression in the scan near pixel 100 corresponding to the site of the repair. This measured depression in transmission corresponded to a loss of only 2%, well within the allowable transmission for a successful repair.

These results demonstrate the full range of abilities of femtosecond light pulses to ablatively remove absorber materials with exquisite precision. But perhaps as importantly, conditions can be found in which an absorbing material can be removed from an underlying transparent material without damage and with spatial resolutions below the diffraction limit.

Figure 12 shows several pictures of the Burlington, Vermont femtosecond photomask repair tool. Figure 12(a) shows the development stage of the tool. At left are the monitor and control electronics. A Staubli robotic arm is seen

(a)

(b)

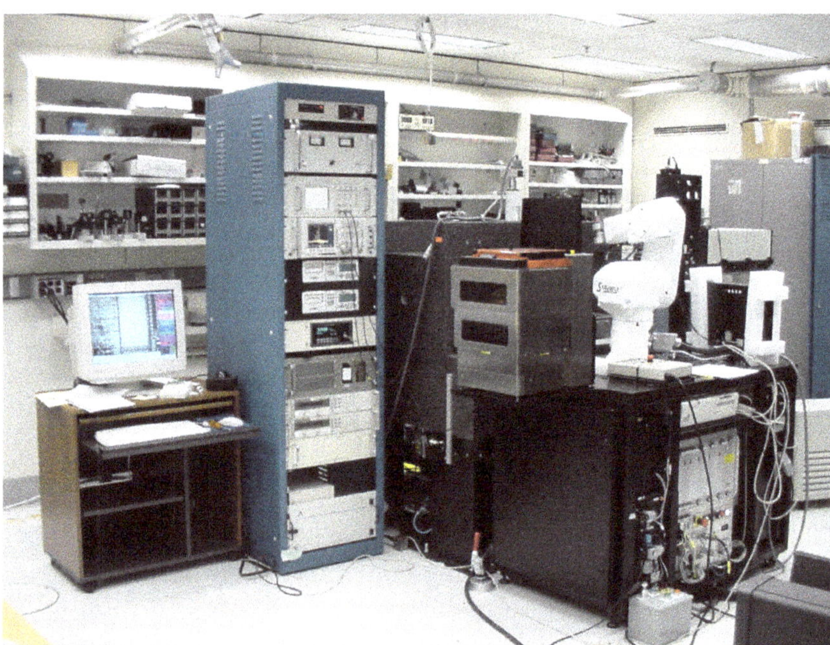

Fig. 12. Pictures of the MARS femtosecond repair tool during construction.

in the foreground and was programmed to load the photomask into the tool without requiring human intervention. A mask loader is seen to the left of the robot. Behind the robot is a large Anorad granite platform with mask positioner. Above the platform, a Leica DUV microscope system is mounted. On the right is a Spectra Physics femtosecond amplified laser system. Control optics are mounted on the granite stage. Figure 12(b) is a different angle view of the system.

The third picture (Fig. 13) shows the MARS femtosecond photomask repair tool installed in the clean room environment of the Burlington, Vermont mask manufacturing facility. The mask loader is seen near the corner of the tool. An operator places a standard mechanical interface (SMIF) pod containing the mask to be repaired into the loader where it is collected by the robotic arm, oriented, and then loaded into the repair tool. All imaging is shown on the monitor to the left. A clean room technician is shown adjusting the tool. The tool is enclosed in a stainless-steel box to convert the laser safety level from class 4 (most dangerous laser) to a class 1 safety level with appropriate interlocks.

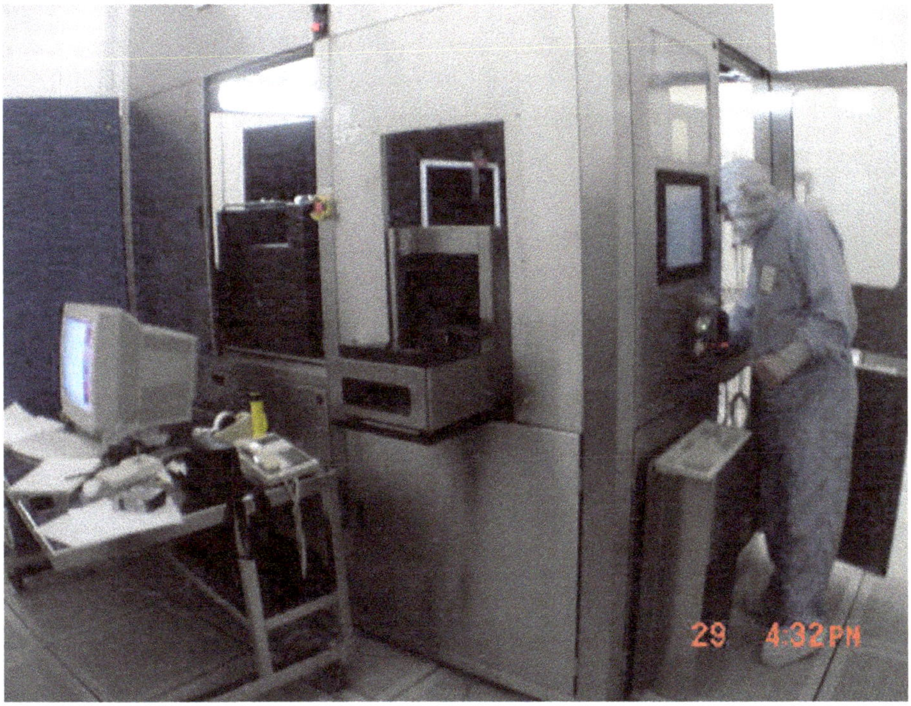

Fig. 13. Technician working on MARS femtosecond repair tool installed in IBM's Burlington, Vt. Mask Manufacturing site.

5.7 Extending Femtosecond Ablation to Shorter Wavelengths

Extending processes, both ablation and perhaps deposition, to shorter wavelengths is desirable from a number of perspectives. Motivated by lithography, complex integrated circuits are presently printed with both 248 nm and 193 nm light. Lithography at 157 nm [17] was in development but appears to have been supplanted by lithography at 13 nm, called extreme ultraviolet (EUV). EUV lithography [18, 27] and repair will be discussed later in this chapter. The advantages of ablating at these wavelengths include improved spatial resolution, which scales linearly with wavelength and the prospects of discovering new physics involving the interaction of intense femtosecond pulses with materials at these wavelengths.

While the advantages are clear, the difficulties associated with shrinking the wavelength are many. The first step would be to both ablate and measure at 193 nm. At present, 193 nm AIMS tools are commercially available but no femtosecond laser repair tools operate at this wavelength. Phase matching in nonlinear crystals becomes difficult below 200 nm and radiation damage to these materials limits their usable lifetime. Obtaining optical components from commercial sources also becomes difficult and costly.

To address the issue of phase matching, sub 200 nm femtosecond pulses can be generated in the gas phase, with sufficient intensity to ablate Cr. In Fig. 14, a schematic setup of a capillary/gas phase system for upconversion of 800 nm light to 200 nm and below is shown. The 800 nm light is frequency doubled to 400 nm, passed through an appropriate delay line and mixed

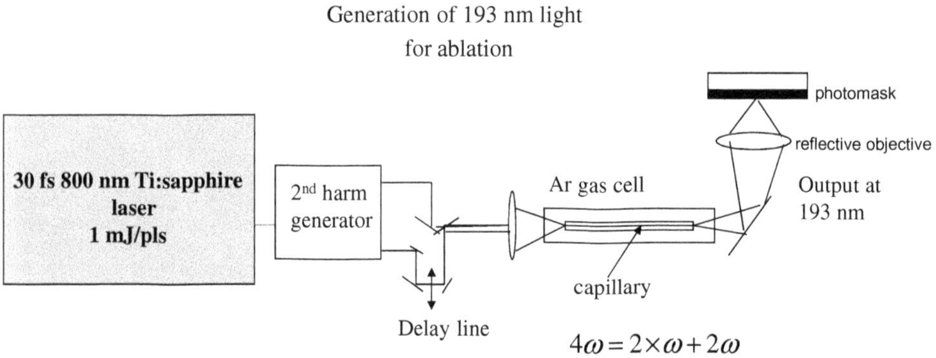

$$4\omega = 2\times\omega + 2\omega$$
$$4\omega = 2\times 3\omega - 2\omega$$
$$4\omega = 2\omega + 3\omega - \omega$$
$$4\omega = 3\times 2\omega - 2\times\omega$$

Fig. 14. Schematic of method to generate 193 nm light with femtosecond pulses using gas phase 4-wave mixing [28].

with 400 nm light in a hollow core glass fiber resident inside an Ar filled cell [28]. The hollow core fiber, or capillary which is 30 cm long, functions as a waveguide that substantially increases the interaction length for converting the input light, via sum and difference frequency mixing, into wavelengths near 266, 200, and 160 nm. By varying the Ar pressure in the cell, and hence within the capillary, phase matching can be optimized. A peak pulse energy of $\sim$0.5 μJ at 200 nm was generated at an Ar pressure of $\sim$400 torr [12], separated from other wavelengths with a series of four mirrors coated for high reflectance at 193 nm. Shorter wavelengths can be achieved by tuning the seed Ti:sapphire oscillator to below 800 nm; for example, tuning the oscillator to 772 nm produces 193 nm light from the capillary output. This light was then focused with a 0.5 NA Schwarzchild reflective objective, coated for high reflectance at 193 nm, onto the surface of a Cr-on-glass photomask. Figure 15 shows a 248 nm transmitted light optical micrograph of a horizontal line ablated in the Cr with 30 fs, 2.8 nJ pulses of 200 nm light. As with ablation using 266 nm, 100 fs pulses, the Cr is completely removed with no observable damage to the underlying fused silica substrate. In the future, higher quality, higher NA objectives will be used to determine the gains in spatial resolution expected at these shorter wavelengths.

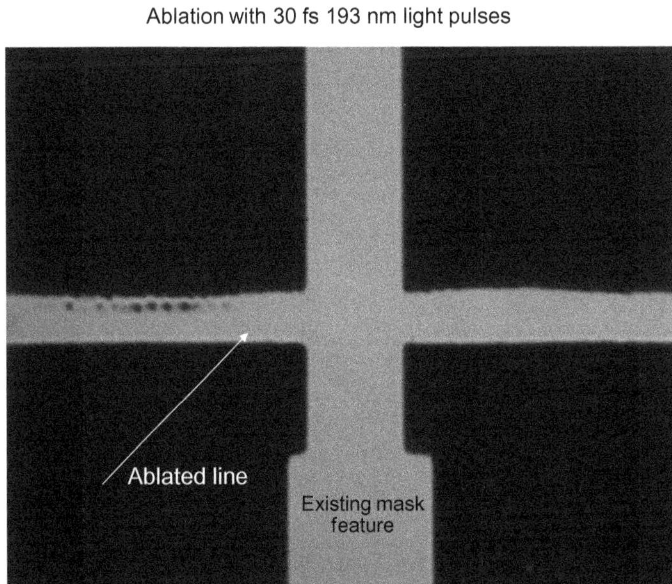

Fig. 15. Transmitted optical micrograph of a line ablated into a chrome-on-glass photomask with 30 fs 193 nm light pulses.

A second perhaps simpler method of generating sub-200 nm pulses of light involves the use of nonlinear crystals. Instead of 4-wave mixing in Ar gas, crystals of beta-barium borate (BBO) can be employed. In this approach, the 800 nm light pulses can be frequency quadrupled in a series of BBO crystals. By tuning the input light below 800 nm, 193 nm light can be generated but the efficiency drops dramatically near the end of the phase matching capability of the crystals.

5.8 Mask Repair in the Soft X-ray Wavelength Regime

With the introduction of EUV lithography and reflective mask technology for EUV, issues of repair remain critically important. EUV lithography operates at a wavelength of 13.5 nm. The standard EUV stepper tool is broken into three key sections, similar to optical and DUV (193 nm) lithography. These sections include light source, demagnification optics and stage. Fundamentally different here are the light source and optics. The light source that has been developed for EUV lithography involves the formation of a plasma by focusing pulses from an intense CO_2 laser onto molten droplets of Sn or other high atomic number metals.

The light from the ensuing plasma is collected and directed to a series of multilayer-coated mirrors. The multilayer mirror substrate [29, 30] is constructed from a stack of alternating layers of Si (low Z) and Mo (high Z) in much the same fashion as dielectric reflection mirrors are constructed for the visible region of the spectrum using layers of transparent materials with varying index of refraction. Typical reflectance from a multilayer mirror optimized for 13.5 nm is close to 70% but since there are as many as six mirrors in the optical train the light intensity is reduced to about 10% of the original input light. Typically, 250 W or more of primary 13.5 nm light must be produced and collected at the source for sufficient light to be available at the wafer surface to expose the photoresist.

While there are many technical challenges to implementing EUV technology, tool development moves forward and repair of these reflective photomasks is required. For such a mask, the multilayer mirror itself is just the reflective substrate and the absorber defining the circuit pattern must be deposited on the mirror in the same fashion as for transmitting photomasks. And, of course, errors can occur. Absorbing defects can be removed by using AFM type tools to scratch off the absorber defect assuming the underlying mirror stack is not affected.

Although a repair tool has not yet been built, it may be possible to generate and select intense pulses of light at 13.5 nm through high-harmonic generation, focused to a small spot by a multilayer mirror microscope objective or Fresnel zone plate [31] to intensities where the absorbing defect is ablatively removed but low enough so that the reflective mirror underneath is not damaged, but instead reflected away by the mirror. For such a strategy to work, a high degree of control of the harmonic flux would be required. If EUV lithography enters into wide use, such at-wavelength (actinic) repair schemes will become increasingly important. Utilizing femtosecond 32 nm XUV pulses from a free electron laser, ablation has been carried out on Si substrates to good effect [32].

5.9 Cr Deposition with Femtosecond Pulses at Atmospheric Pressure

While in the predominant defect on photomasks is typically excess absorber, missing absorber material also presents a significant problem and we now discuss how to add material, specifically Cr to locations where it is missing [33] using 100 fs pulses of light at atmospheric pressure.

Deposition of Cr is shown schematically in Fig. 16. Femtosecond pulses of light are focused through a saturated atmosphere of $Cr(CO)_6$. The saturated atmosphere produces a physically adsorbed layer of organometallic. By focusing an intense femtosecond pulse of light into this layer, photodecomposition results. While it would seem reasonable to use UV pulses at, for example, 266 nm to deposit Cr since typically short wavelength light readily photodissociates organometallic compounds and provides potentially smaller spot sizes, the cross-section for this process is too high and decomposition occurs

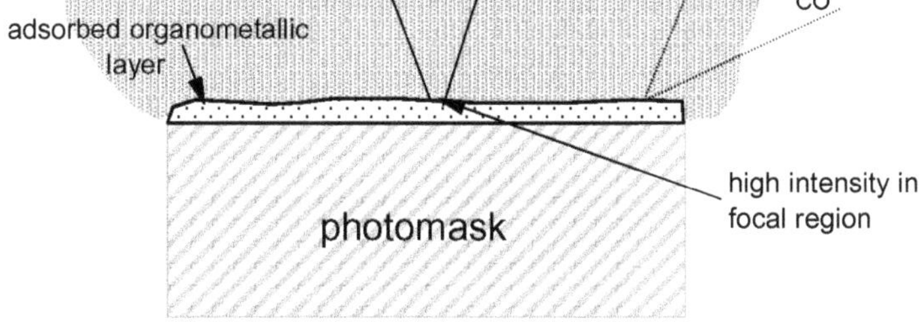

Fig. 16. Schematic of femtosecond gas phase metal deposition system.

in the gas phase above the substrate. As a consequence, in addition to the desired spatially localized deposition of Cr metal, a good deal of Cr metal haze is observed due to the gas phase decomposition component.

This problem is avoided by using photons whose wavelength corresponds to a smaller organometallic dissociation cross-section at low intensities — in the linear regime. But at higher intensities at and near the focus, nonlinear processes such as two-photon absorption are then utilized to dissociate the adsorbed organometallic compound while enhancing the spatial localization of the deposit.

This was achieved through irradiation with 400 nm, 120 fs pulses [33]. At this wavelength, the $Cr(CO)_6$ is essentially transparent to the light, since its absorption cross-section at 400 nm is $\leq 3 \times 10^{-20}\,\text{cm}^2$. At the focus, the high pulse intensity drives decomposition of the organometallic. Figure 17 shows a schematic of the experimental setup. The deposition system consists of a microscope with an objective optimized for transmission of 400 nm, femtosecond light pulses. Surrounding the microscope objective is a jacket that transports the compound from the source to the region between the objective and substrate. Nitrogen gas flows through a charge of $Cr(CO)_6$ captured between two glass frits. The saturated gas flows through a jacket surrounding a 0.9 NA (Leica) objective that images at 365 nm, creating a laminar flow of gas under the objective. The 400 nm light pulses are focused at the surface of the glass substrate and a computer-controlled piezoelectric stage scans the substrate under the objective.

Figure 18 shows optical micrographs taken in transmission with 248 nm light and in reflection at 365 nm of the deposited Cr lines. Also shown is an

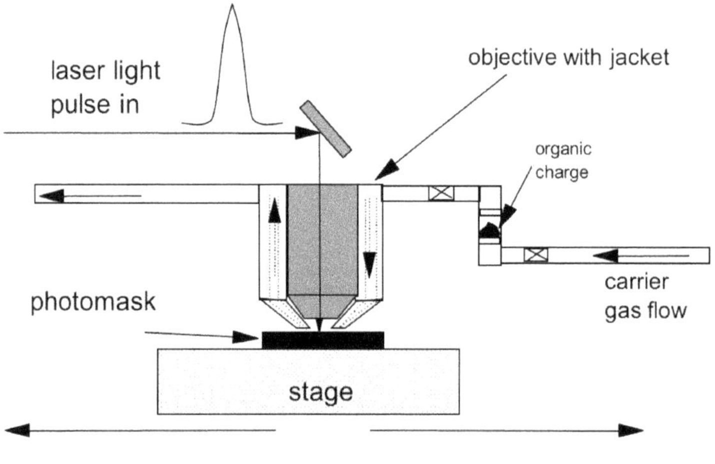

Fig. 17. Schematic of 400 nm fs Cr deposition.

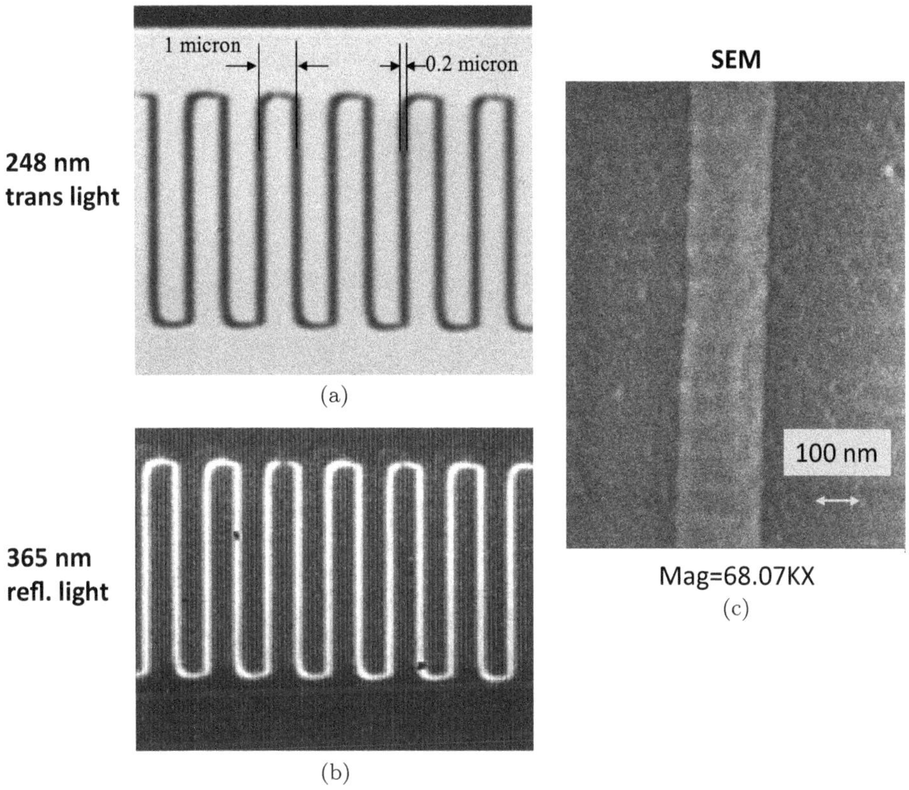

Fig. 18. (a) 248 nm transmission optical micrograph of deposited Cr lines indicating that they are optically opaque. (b) 365 nm reflective optical micrograph of Cr lines showing that they are smooth and reflective. (c) SEM micrograph of a 100 nm Cr line.

SEM of a Cr line. The 248 nm transmission micrograph exhibits deposited, opaque Cr lines of 200 nm width. The 365 nm micrograph captures the reflective nature of the deposited lines that show up brightly, indicating they are metallic and smooth as confirmed by the SEM. No Cr haze or overall deposition is observed in between the deposited lines. The linewidth is 200 nm which is close to but slightly smaller than the $\lambda/2$ NA value calculated for 400 nm light pulses.

A scanning nanoprofilometer was used to measure the height of the deposited lines and those results are displayed in Fig. 19. The height of the Cr lines are measured to be 40–45 nm with sharp sidewalls and no deposited material observed in between. The linewidths, are 200 nm after deconvolving the profilometer tip contribution to the measurement, in agreement with the optical and SEM micrographs and the deposits adhere well to the substrate-fused silica mask. Aggressive cleaning with nitric acid did

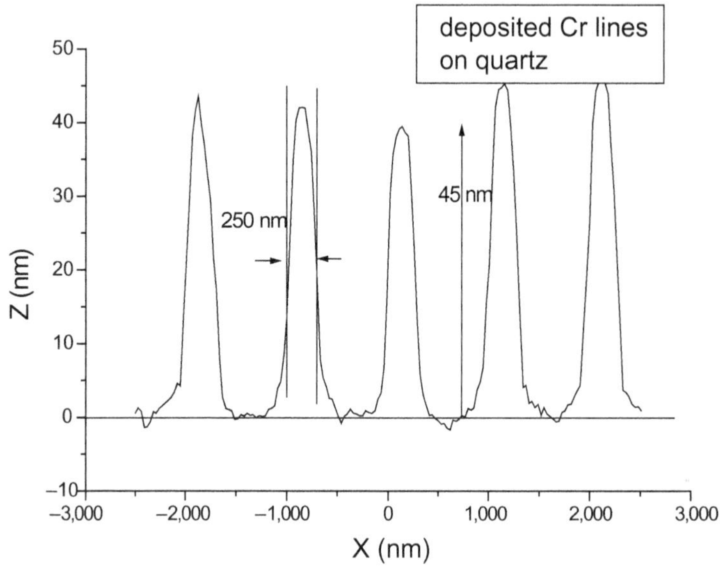

Fig. 19. Profilometry of Cr lines showing a height of 40–45 nm and deconvolved width of 200 nm.

not remove the deposits — such acid treatments are typically used in the post-repair cleaning of photomasks.

The appearance of the deposits on the fused silica substrates depends strongly on the intensity of incident 400 nm light and this is an indication that nonlinear photochemical processes are at play here. A series of measurements was carried out in which the minimum average power required to observe the onset of deposition was recorded as a function of the laser repetition rate, from 10 kHz to 300 kHz and are displayed in Fig. 20a.

Note the linear relationship between the laser power and repetition rate; when the repetition rate is divided out the threshold intensity required to form an observable deposit is $7.8 \pm 1 \times 10^{11}$ W/cm^2 (Fig. 20b). The intensity dependent nature of the deposition process indicates a multiphoton, photolytic process driven by multiphoton dissociation of the molecules adsorbed onto the fused silica substrate. This nonlinear behavior is critically important as it indicates that photodissociation will occur only in the focal region and not in the gas phase. In other words, deposition occurs only where it is desired.

Increasing the thickness of the deposited Cr is beneficial as it eliminates any residual light transmission through thinner deposits. Multiple scans were performed and are shown in Fig. 21, where an increasingly roughened deposit

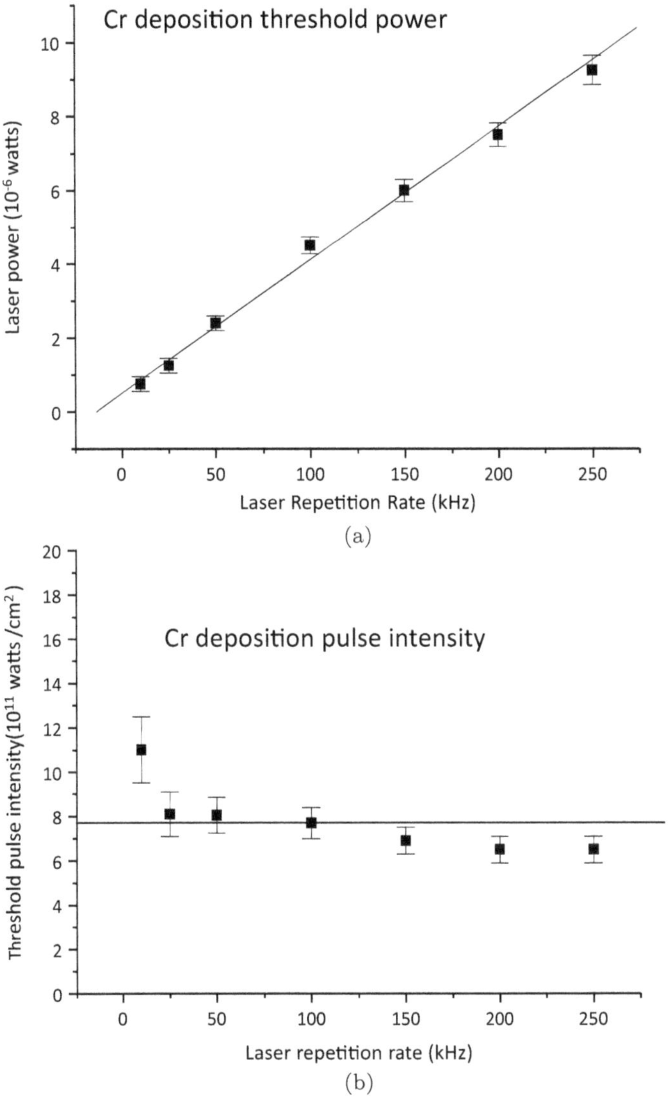

Fig. 20. (a) Threshold power at which Cr deposits form on fused silica as a function of laser repetition rate. (b) Threshold pulse intensity at which Cr deposits form on fused silica as a function of laser repetition rate.

is observed. Dark patches, starting as small spots, were observed to grow with successive scans until finally the entire serpentine darkened.

Figure 22 shows in detail, through an optical reflection micrograph and SEM, that a roughened deposit has formed. In addition, we note in the reflection micrograph, a haze extending roughly 1 micron beyond the lines

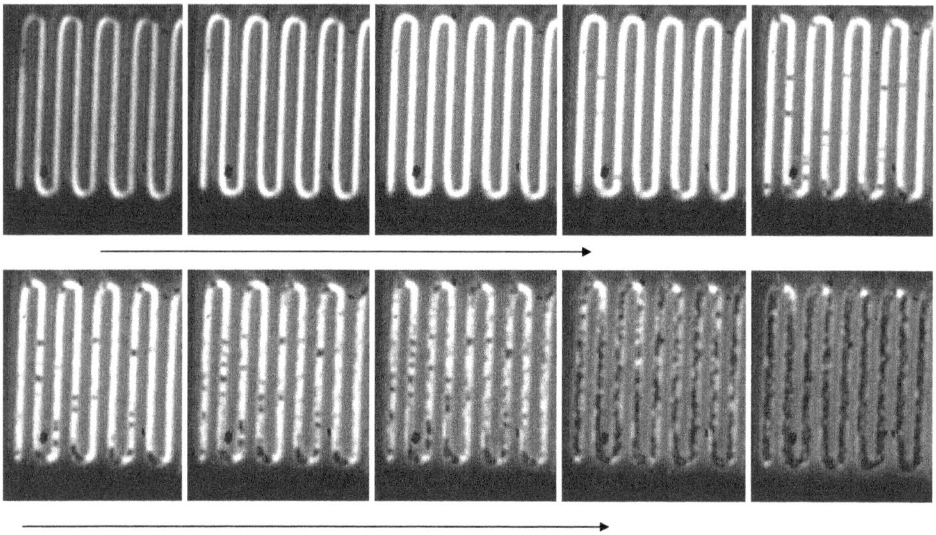

Fig. 21. Successive femtosecond-laser deposition scans showing increasing roughness.

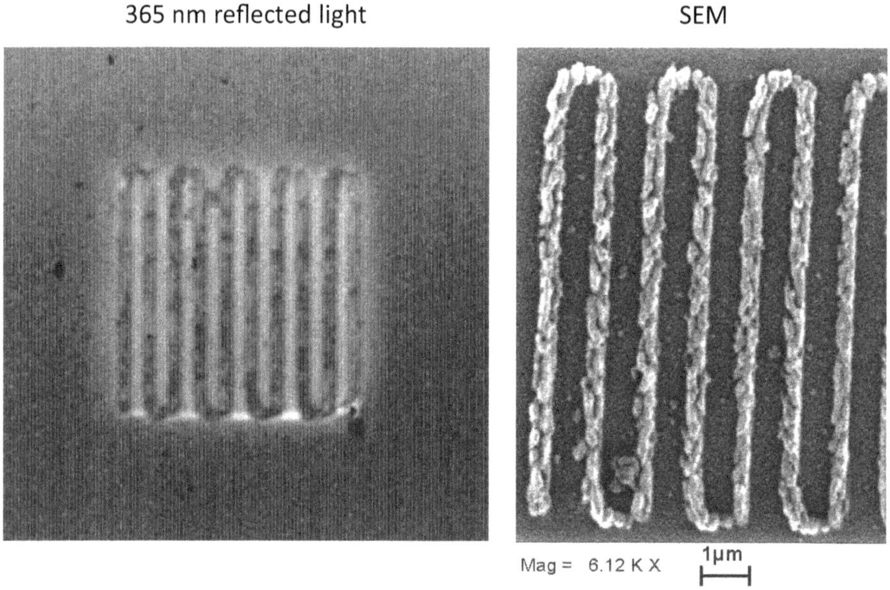

(a)

(b)

Fig. 22. (a) 365 nm optical micrograph of multiple Cr deposition scans leading to roughened deposits. (b) SEM of scans revealing the roughened deposits.

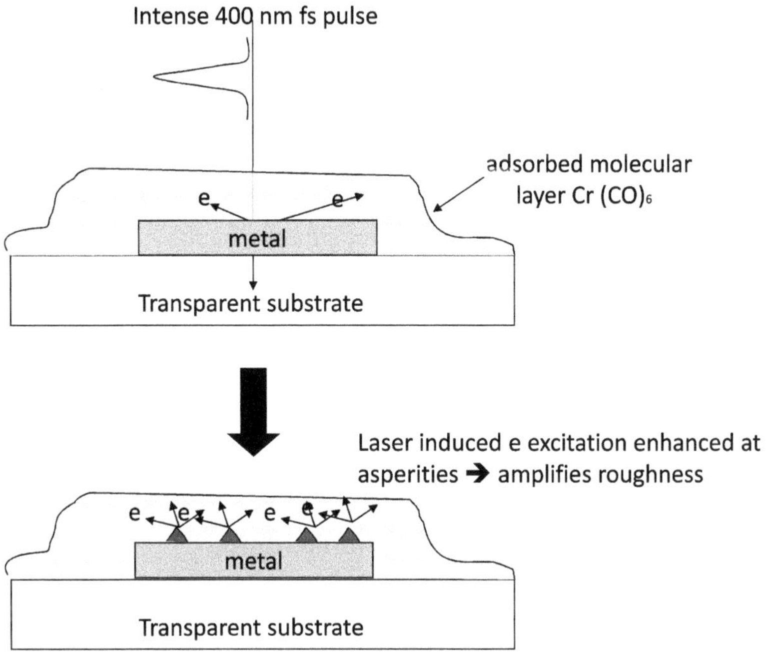

Fig. 23. Schematic model for multiphoton electron emission giving rise to roughened deposits and haze in adjacent areas of the mask.

is observed. While the initial Cr deposit is formed on glass, a non-absorbing substrate, subsequent scans deposit Cr on existing Cr lines.

When intense femtosecond pulses of 400 nm light impinge upon metals, substantial fluxes of electrons are liberated due to multiphoton electron emission (Fig. 23). The electron emission flux was measured to be $3 \times 10^{16}/\mathrm{cm}^2\mathrm{s}$ for 400 nm pulses with intensities of $\sim 8 \times 10^{11}\,\mathrm{W/cm}^2$, the threshold for metal deposition on glass, as described earlier.

Studies of electron beam stimulated dissociation of organometallics [34], such as $(CH_3)Al$ adsorbed on semiconductor surfaces have found cross-sections of $0.7 - 2 \times 10^{-16}\,\mathrm{cm}^2$ for electrons with energies from 10 eV to 50 eV. In addition, the gas phase mean-free path of electrons with energies in the 10 eV to 50 eV range is ~1 micron, and hence the haze observed is attributable to electron stimulated decomposition in the gas phase in the region of the deposited metal lines.

The line roughening with subsequent deposition scans may be postulated to be associated with metal asperities which form during the deposition process. Any roughness, however minute is amplified in additional scans which deposit new layers since multiphoton electron emission is strongly enhanced on roughened surfaces due to concentrated electric fields [35] at the tips of the deposited material.

Metal deposition with femtosecond light pulses can be described as follows. On transparent materials, multiphoton absorption within the adsorbed molecular layers drives decomposition resulting in metal deposition. The substrate plays only a passive role, allowing the adsorption of the precursor compound. In contrast, on absorbing substrates the intense femtosecond pulses produce high fluxes of electrons through multiphoton emission; these energetic electrons drive dissociation within the adsorbed molecular layer leading to additional, but increasingly roughened metal deposits. Nonetheless, deposition with femtosecond light pulses produces deposited metal in a well-defined manner suitable for many applications requiring a high resolution process.

5.10 Survey of Femtosecond Micromachining

Micromachining with femtosecond lasers spans a wide array of applications that would, by itself, fill an entire book. To address this conundrum, we describe several additional applications. The simplest use of femtosecond ablation is to drill clean, precise holes as shown in Fig. 24. As was observed in femtosecond mask repair, minimal heat transfer produces holes with crisp edges while nanosecond ablation produces melting, rolled edges, and splatter. The lack of melted metal along with minimal shock propagation to the adjacent material produces extremely sharp ablated patterns.

This effect can be extended to numerous materials that are sensitive to issue of shock propagation. One example is in dentistry where nanosecond lasers used to remove decay (cavities) resulted in adjacent cracking of the dental material due to shock propagation while femtosecond pulses showed no adjacent cracks, [36–38].

Femtosecond lasers have been used to replace scalpels in eye surgery. Excimer reshaping of the cornea to improve vision is known as LASIK or laser assisted *in situ* keratomileusis. In this surgical procedure, the opthalmologist uses either a microkeratome (a mechanical surgical tool) to create a thin circular flap in the cornea. The flap is folded back to reveal the middle layer of cornea called the stroma. The microkeratome has been in some cases replaced with a femtosecond laser where pulses are used to cut the corneal tissue with minimal damage, creating the thin flap. Excimer laser pulses, strongly absorbed in the cornea are then used to ablatively reshape the cornea itself providing the optical correction and sharper vision.

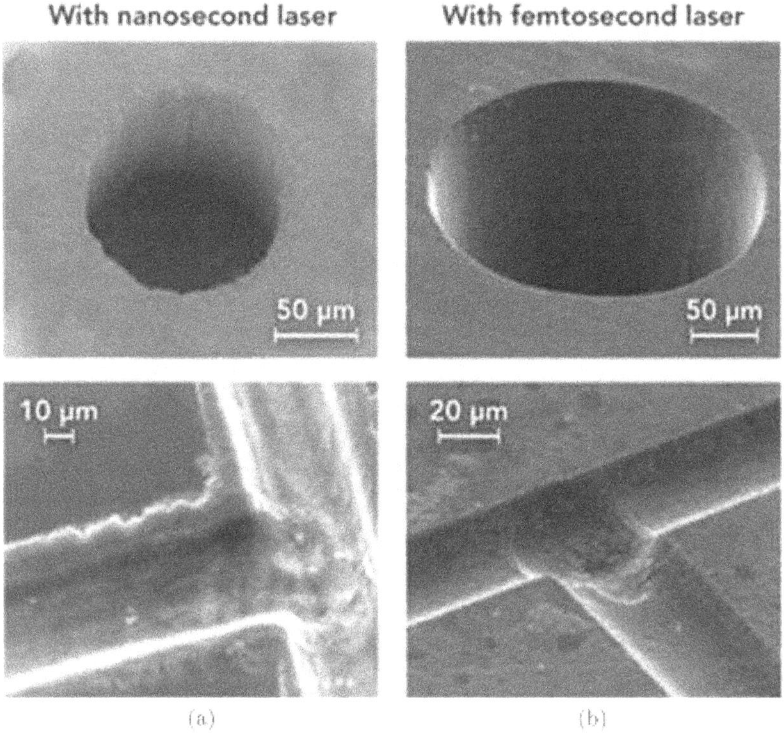

Fig. 24. Holes and trenches ablated with 266 nm nanosecond laser pulses (a) and 800 nm, 100 fs laser pulses (b) in glass.

5.11 Femtosecond Micromachining of Glasses and Transparent Materials

While materials such as metals and semiconductors absorb individual visible light pulses rather strongly, transparent materials do not. Consider the example of fused silica that we have been discussing extensively in this chapter. It is the substrate upon which absorbing materials are deposited in the formation of a photomask. And because the metal absorber atop the fused silica substrate absorbs individual laser photons while the substrate does not, this provides a contrast in the absorption process that allows for removal of metal defects as we have discussed.

But what can be done when we wish to modify or ablate transparent materials? This has been studied extensively in the literature [9]. The process of modifying or ablating a transparent material with a femtosecond laser pulse is fundamentally different than for an opaque material (Fig. 25).

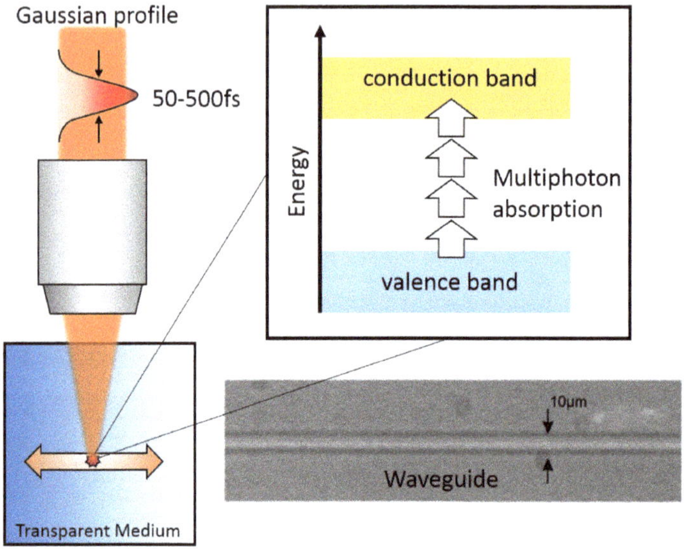

Fig. 25. Schematic showing focusing femtosecond light pulses with a microscope objective into a transparent material, multiphoton absorption promotes electrons from the valence to conduction bands of the glass changing the refractive index of the material. When scanned compaction leads to the formation of a waveguide within the glass.

In order for a transparent medium to absorb light either very short wavelength photons that exceed the band gap energy or an intense pulse of visible (sub-band gap) light leading to multiphoton absorption must be employed. The latter is shown in Fig. 25. Let us consider SiO_2. It is a very wide band gap insulating material that is found in many forms depending upon the impurities within the material itself. Common glasses such as soda–lime glass is SiO_2 with impurities that included Na (hence "soda" in the name) and is truly a glass in that it is a highly disordered solid with only very short range order. This material begins to absorb at ~350 nm so it is transparent to the eye but does not transmit in the UV. Removal of the impurities changes the properties of the material and widens the band gap. In the absence of impurities SiO_2 will crystallize. The amorphous form is called fused silica and has a band gap of ~7.5 eV. This corresponds to a wavelength of 165 nm. Generating a high intensity femtosecond pulse of 7.5 eV light is challenging; there are no commercially available nonlinear crystals that can produce this wavelength. It may be possible to reach this wavelength via four-wave mixing as discussed in Section 5.7 or via high-harmonic generation in gases.

A second approach is to use visible light of sufficient intensity so that multiphoton absorption occurs [9, 39, 40] as shown in Fig. 25. By focusing a femtosecond pulse into a transparent medium such as glass, an extremely small volume delineated by the confocal parameter of the focused light within the glass experiences an extraordinarily high peak pulse intensity.

Multiphoton absorption and avalanche ionization leads to the substantial modification of the material within the focal region. Outside of the focal region the intensity drops dramatically and material changes subside rapidly. By scanning the laser or material as well as changing the depth at which the laser focuses, three-dimensional machining of the glass can be accomplished.

At modest intensities, the refractive index of the material can be modified through a process called compaction in which exposure to UV light or femtosecond light pulses causes a densification of the glass that increases its index of refraction. By changing the index of refraction in a localized region below the surface of the glass and then scanning along a linear path, a waveguide can be created. This is accomplished by increasing, through compaction, the index of refraction in the irradiated region, scanned to create a cylindrical tube in the overall medium. Since the index within the irradiated tube is higher than the surrounding region, propagating light is trapped due to total internal reflection and waveguiding is achieved; the propagating light is then characterized by a transverse mode [9, 41].

In another compelling application femtosecond micromachining can be used to create microfluidic channels within glass and other materials [42]. Through the process of multiphoton absorption (as shown in Fig. 26) free electrons promoted into the conduction band of the glass are also accelerated by the high electric field of the laser, similar to the physical processes in high-harmonic generation. These excited electrons induce avalanche ionization

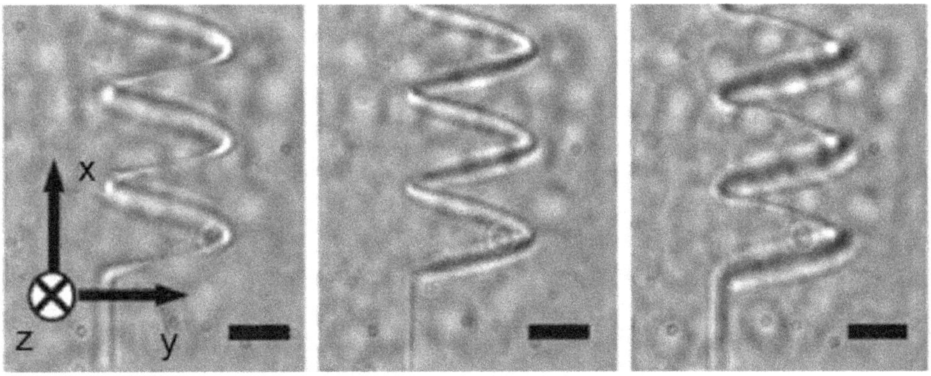

Fig. 26. Optical micrographs of spiral microfluidic channels fabricated in a transparent polymer block created through a combination of two-photon polymerization and multiphoton ablation (after Ref. [42]).

and generate a plasma in the focal region that leads to the creation of a void [23].

Shown in Fig. 26 are spiral microfluidic channels machined in a polymer by using a combination of two-photon polymerization and multiphoton ablation.

As can be seen in these examples, femtosecond laser ablation is an effective tool for machining a wide range of materials including metals, ceramics, and glasses. Extraordinarily high spatial resolution can be achieved, well into the sub-micron and even nanometer regime. From the semiconductor industry to biomedical devices, optical devices including waveguides as we have discussed can all be improved by machining with femtosecond laser pulses.

References

[1] M. Bass, *Laser Materials Processing*, vol. 3. Elsevier, Amsterdam, 2012.
[2] J. Poate, *Laser Annealing of Semiconductors*, Elsevier, Amsterdam, 2012.
[3] T. P. Hughes, "Plasmas and laser light," *Halsted Press, New York*, vol. 1, 539 p., 1975.
[4] W. S. Fann, R. Storz, H. W. K. Tom, and J. Bokor, "Electron thermalization in gold," *Phys. Rev. B*, vol. 46, no. 20, p. 13592, 1992.
[5] M. Hauer, D. J. Funk, T. Lippert, and A. Wokaun, "Time resolved study of the laser ablation induced shockwave," *Thin Solid Films*, vol. 453, pp. 584–588, 2004.
[6] X. Liu, D. Du, and G. Mourou, "Laser ablation and micromachining with ultrashort laser pulses," *IEEE J. Quantum Electron.*, vol. 33, no. 10, pp. 1706–1716, 1997.
[7] P. P. Pronko, S. K. Dutta, J. Squier, J. V Rudd, D. Du, and G. Mourou, "Machining of sub-micron holes using a femtosecond laser at 800 nm," *Opt. Commun.*, vol. 114, no. 1, pp. 106–110, 1995.
[8] M. Lenzner, J. Krüger, S. Sartania, Z. Cheng, C. Spielmann, G. Mourou, W. Kautek, and F. Krausz, "Femtosecond optical breakdown in dielectrics," *Phys. Rev. Lett.*, vol. 80, no. 18, p. 4076, 1998.
[9] R. R. Gattass and E. Mazur, "Femtosecond laser micromachining in transparent materials," *Nat. Photonics*, vol. 2, no. 4, pp. 219–225, 2008.
[10] E. G. Gamaly, A. V Rode, B. Luther-Davies, and V. T. Tikhonchuk, "Ablation of solids by femtosecond lasers: Ablation mechanism and ablation thresholds for metals and dielectrics," *Phys. Plasmas*, vol. 9, no. 3, pp. 949–957, 2002.
[11] B. K. Ridley, *Quantum Processes in Semiconductors*. Oxford University Press, Oxford, 2013.
[12] R. Haight, A. Wagner, P. Longo, and D. Lim, "High resolution material ablation and deposition with femtosecond lasers and applications to photomask repair," *J. Mod. Opt.*, vol. 51, no. 16–18, pp. 2781–2796, 2004.
[13] S. A. Self, "Focusing of spherical Gaussian beams," *Appl. Opt.*, vol. 22, no. 5, pp. 658–661, 1983.
[14] W. Vollrath, "Ultra-high-resolution DUV microscope optics for semiconductor applications," in *Proc. SPIE*, 2005, vol. 5865, pp. E1–9.
[15] H. J. Levinson, "Principles of lithography," 2005.

[16] B. W. Smith, A. Bourov, H. Kang, F. Cropanese, Y. Fan, N. Lafferty, and L. Zavyalova, "Water immersion optical lithography at 193 nm," *J. Micro/Nanolithography, MEMS, MOEMS*, vol. 3, no. 1, pp. 44–51, 2004.
[17] M. Switkes and M. Rothschild, "Immersion lithography at 157 nm," *J. Vac. Sci. Technol. B*, vol. 19, no. 6, pp. 2353–2356, 2001.
[18] V. Bakshi, *EUV lithography*, vol. 178. Spie Press, Bellingham, 2009.
[19] A. K.-K. Wong, *Resolution Enhancement Techniques in Optical Lithography*, vol. 47. SPIE press, Bellingham, 2001.
[20] R. Haight and J. A. Silberman, "Surface intervalley scattering on GaAs (110): Direct observation with picosecond laser photoemission," *Phys. Rev. Lett.*, vol. 62, no. 7, p. 815, 1989.
[21] R. R. Alfano, *Semiconductors Probed by Ultrafast Laser Spectroscopy*. Elsevier, Amsterdam, 2012.
[22] J. Shah, "Ultrafast studies of carrier relaxation in semiconductors and their microstructures," *Superlattices Microstruct.*, vol. 6, no. 3, pp. 293–302, 1989.
[23] B. N. Chichkov, C. Momma, S. Nolte, F. Von Alvensleben, and A. Tünnermann, "Femtosecond, picosecond and nanosecond laser ablation of solids," *Appl. Phys. A*, vol. 63, no. 2, pp. 109–115, 1996.
[24] R. Haight, P. Longo, and A. Wagner, "Material processing using femtosecond lasers?: repairing photomasks," *MRS Bull.*, vol. 31, no. August, pp. 634–638, 2006.
[25] R. Haight, D. Hayden, P. Longo, T. E. Neary, and A. Wagner, "Implementation and performance of a femtosecond laser mask repair system in manufacturing," in *18th Annual BACUS Symposium on Photomask Technology and Management*, 1998, pp. 477–484.
[26] R. A. Budd, D. B. Dove, J. L. Staples, R. M. Martino, R. A. Ferguson, and J. T. Weed, "Development and application of a new tool for lithographic mask evaluation, the stepper equivalent Aerial Image Measurement System, AIMS," *IBM J. Res. Dev.*, vol. 41, no. 1.2, pp. 119–129, 1997.
[27] B. Wu and A. Kumar, "Extreme ultraviolet lithography: A review," *J. Vac. Sci. Technol. B*, vol. 25, no. 6, pp. 1743–1761, 2007.
[28] L. Misoguti, S. Backus, C. G. Durfee, R. Bartels, M. M. Murnane, and H. C. Kapteyn, "Generation of broadband VUV light using third-order cascaded processes," *Phys. Rev. Lett.*, vol. 87, no. 1, p. 13601, 2001.
[29] E. Spiller, *Soft X-ray Optics*. SPIE Optical Engineering Press, Bellingham, WA, 1994.
[30] E. Spiller, A. Segmüller, J. Rife, and R. Haelbich, "Controlled fabrication of multilayer soft-X-ray mirrors," *Appl. Phys. Lett.*, vol. 37, no. 11, pp. 1048–1050, 1980.
[31] G. Vaschenko, C. Brewer, F. Brizuela, Y. Wang, M. A. Larotonda, B. M. Luther, M. C. Marconi, J. J. Rocca, C. S. Menoni, and E. H. Anderson, "Sub-38 nm resolution tabletop microscopy with 13 nm wavelength laser light," *Opt. Lett.*, vol. 31, no. 9, pp. 1214–1216, 2006.
[32] N. Stojanovic, D. Von der Linde, K. Sokolowski-Tinten, U. Zastrau, F. Perner, E. Förster, R. Sobierajski, R. Nietubyc, M. Jurek, and D. Klinger, "Ablation of solids using a femtosecond extreme ultraviolet free electron laser," *Appl. Phys. Lett.*, vol. 89, no. 24, p. 241909, 2006.
[33] R. Haight, P. Longo, and A. Wagner, "Metal deposition with femtosecond light pulses at atmospheric pressure," *J. Vac. Sci. Technol. A*, vol. 21, no. 3, pp. 649–652, 2003.
[34] C. P. A. Mulcahy, J. Eggeling, and T. S. Jones, "Low-energy electron beam induced dissociation of methyl groups chemisorbed on semiconductor surfaces: (CH 3) 3 Al adsorbed on GaAs and InSb," *Chem. Phys. Lett.*, vol. 288, no. 2, pp. 203–208, 1998.

[35] A. Nitzan and L. E. Brus, "Theoretical model for enhanced photochemistry on rough surfaces," *J. Chem. Phys.*, vol. 75, no. 5, pp. 2205–2214, 1981.
[36] J. Serbin, T. Bauer, C. Fallnich, A. Kasenbacher, and W. H. Arnold, "Femtosecond lasers as novel tool in dental surgery," *Appl. Surf. Sci.*, vol. 197, pp. 737–740, 2002.
[37] J. Krüger, W. Kautek, and H. Newesely, "Femtosecond-pulse laser ablation of dental hydroxyapatite and single-crystalline fluoroapatite," *Appl. Phys. A Mater. Sci. Process.*, vol. 69, no. 7, pp. S403–S407, 1999.
[38] R. F. Z. Lizarelli, M. M. Costa, E. Carvalho-Filho, F. D. Nunes, and V. S. Bagnato, "Selective ablation of dental enamel and dentin using femtosecond laser pulses," *Laser Phys. Lett.*, vol. 5, no. 1, p. 63, 2007.
[39] A. M. Kowalevicz, V. Sharma, E. P. Ippen, J. G. Fujimoto, and K. Minoshima, "Three-dimensional photonic devices fabricated in glass by use of a femtosecond laser oscillator," *Opt. Lett.*, vol. 30, no. 9, pp. 1060–1062, 2005.
[40] M. S. Giridhar, K. Seong, A. Schülzgen, P. Khulbe, N. Peyghambarian, and M. Mansuripur, "Femtosecond pulsed laser micromachining of glass substrates with application to microfluidic devices," *Appl. Opt.*, vol. 43, no. 23, pp. 4584–4589, 2004.
[41] L. J. Jiang, S. Maruo, R. Osellame, W. Xiong, J. H. Campbell, and Y. F. Lu, "Femtosecond laser direct writing in transparent materials based on nonlinear absorption," *MRS Bull.*, vol. 41, no. 12, pp. 975–983, 2016.
[42] W. Xiong, Y. S. Zhou, X. N. He, Y. Gao, M. Mahjouri-Samani, L. Jiang, T. Baldacchini, and Y. F. Lu, "Simultaneous additive and subtractive three-dimensional nanofabrication using integrated two-photon polymerization and multiphoton ablation," *Light Sci. Appl.*, vol. 1, no. 4, p. e6, 2012.

Chapter 6

Femtosecond Photovoltage Spectroscopy and Device Physics

6.1 Introduction

Ultraviolet photoelectron spectroscopy (UPS) has been used extensively for many years to study the fundamental bulk and surface electronic structure of materials. Inherently a surface sensitive technique, the advent of ultrafast pump probe laser spectroscopy has opened new vistas of measurement opportunity. In this chapter, we describe new applications that expand this time-honored, venerable scientific tool to studies of industrially relevant materials and structures through the merger with femtosecond timescale pump-probe techniques. These investigations include metal–oxide–semiconductor (MOS) thin-film stacks, photovoltaic film stack structures, metal-semiconductor Schottky barrier systems, homo- and heterojunction couples, and more.

A key application, that will be a focus of this chapter, involves a rather old technique called photovoltage spectroscopy [1]. Dating back many decades [2], it was recognized that optical illumination can result in changes in the band bending in semiconductors. A good review of the concepts involved in generating and utilizing a photovoltage in semiconductor systems has been given by Kronik and Shapira [3]. Until the advent of pulsed lasers, illumination was typically accomplished with the use of lamps whose output was modulated by a mechanical chopper. Under such weak illumination only meV changes in band bending were introduced and modeling of the optically excited electron-hole (e-h) population was required. Starting from the pioneering work of Brattain and Bardeen in the early 1950s [1], photovoltage spectroscopy has been used in the last six decades as an extensive source of surface and bulk information on various semiconductors and semiconductor interfaces.

Femtosecond photovoltage spectroscopy (FPS) [4] is a new twist on the technique that exploits electrostatics combined with the unique ability of photoelectron spectroscopy to extract the electronic structure of materials. Materials systems can range from single crystal and polycrystalline semiconductors to amorphous films and thin insulating materials deposited on more conductive materials. This encompasses a wide range of materials employed in semiconductor and photovoltaic devices that are of considerable industrial importance.

6.2 Band Bending from Charged Surface States

We start with the most basic situations [5] and will increase complexity as the chapter progresses. First, consider the energetics of the electronic structure of a doped semiconductor near its surface (Fig. 1). For a perfectly

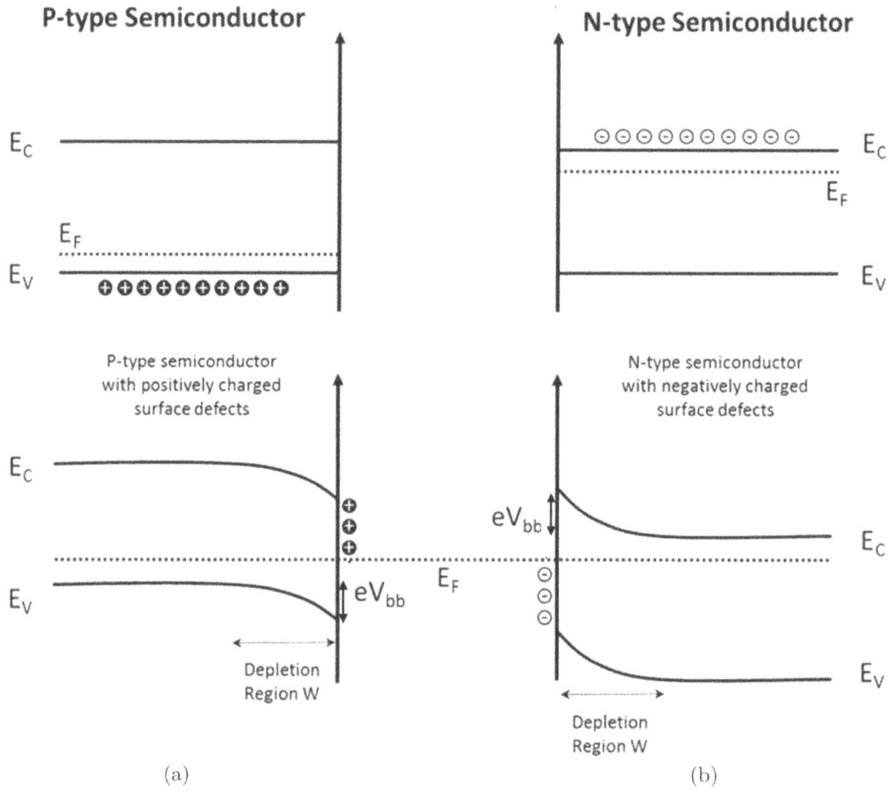

Fig. 1. Electronic structure for p-type (a) and n-type (b) semiconductors. In the absence of surface states (top) the valence (E_V) and conduction (E_C) bands are flat (bottom) introduction of states within the band gap at surface trap charge leading to downward band bending (p-type) or upward band bending (n-type) by an energy eV_{bb}.

terminated semiconductor with no charged states at its surface, the semiconductor bands are essentially flat. For an n-doped semiconductor, negative charge (electrons) will remain in the conduction band as the band gap is devoid of states. For a p-doped semiconductor holes will remain in the valence band.

However, in most semiconducting systems this is not the case. A Si surface will typically have unsaturated (dangling) bonds at the surface and these surface states exist within the band gap of the Si at the surface. For n-doped Si as mentioned above, delocalized electrons near the surface will become trapped in these states and the surface will become negatively charged. With this negative charge at the surface, a dipole field is formed between the trapped charge and the semiconductor. Delocalized electrons within the semiconductor will be repelled by the trapped surface charge and driven into the interior of the semiconductor. The bands bend upward, creating a region depleted of free electrons (W). Screening of the dipole field consistent with the dielectric properties of the semiconductor (with dielectric constant (ε) determines the spatial extent of the depletion region, W. The same argument applies to a p-doped semiconductor where the surface becomes positively charged, is attractive to electrons and the bands bend downward.

Consider an n-type semiconductor such arsenic-doped Si. Since As possesses an additional electron relative to Si (i.e. it is just to the right of Si in the periodic table), when As substitutes for a Si atom in the bulk of the crystal (substitutional site) it becomes ionized, donating its additional electron (relative to Si) to the conduction band of Si. Hence As is a donor atom. The majority carriers, electrons in this case, are delocalized within the bulk of the Si crystal. Typical dopant densities can range from $\sim 10^{15}$ to $10^{21}/\text{cm}^3$ and are limited by the solubility of the dopant in the semiconductor host. Once doped, the high concentration of electrons in the conduction band shifts the Fermi level of the semiconductor toward the conduction band, lowering its work function.

For a p-type semiconductor, the situation is essentially the same as described above but reversed. For B-doped Si, the one fewer valence electron of B results in a delocalized population of holes within the valence band. B is an acceptor in Si. The Fermi level is now found near the valence band maximum and the work function is higher.

In an n-type semiconductor the donated electrons are delocalized within the bulk and those that venture near the surface can be captured by defect states within the band gap of the semiconductor at the surface. Any state occupied with an electron at the surface must be below the Fermi level which for our n-type semiconductor is near the conduction band minimum. Trapping of the electron at the surface then leaves the surface negatively charged. A semiconductor possesses a dielectric constant that lies in the range intermediate between an insulator and a metal and as a result, the electrostatic field is screened over a depth dependent upon both the doping density and the dielectric constant of the material. This dielectric screening results in a quadratic decay in the magnitude of the shift of the bands from the surface into the bulk and gives rise to the term "band bending" to describe this behavior. More detailed discussions are found in the classic text by S.M. Sze (Physics of Semiconductor Devices [5]).

Bending of the bands near the surface occurs when otherwise empty surface states become charged. There are a variety of situations in which this may occur such as lack of termination of surface atoms (dangling bonds), impurity states, defects in the form of missing surface atoms, antisite defects in multi-elemental materials and more.

6.3 Solution of Poisson's Equation

Charge in states localized at the surface induce a dipole field within the semiconductor. A universal aspect of this field is that it can be formed in a wide array of structures that involve the separation of positive and negative charges. In addition to charged surface and/or defect states similar dipole fields can be formed in metal-semiconductor couples that form a Schottky barrier, in cases where an oxide is deposited on a semiconductor where the oxide itself may possess trapped charges, in semiconductor/semiconductor heterostructures, and in photovoltaic stacks, to name just a few.

In photovoltaic stacks, the separation of charge in a doped P–N junction can be treated in a similar fashion as that for heterojunctions as well as metal–semiconductor systems. The equilibration of the Fermi levels in the two contacted systems results in the transfer of charge. Charge transfer in these systems and the resulting impact on the semiconductor bands can be calculated by solving for the potential as a function of distance within the semiconductor itself. As in most cases we are dealing with films extended in two dimensions and we need only solve the one-dimensional Poisson's equation for the potential as a function of distance within the semiconductor. Consider a typical example, that of a metal contact on a semiconductor whose doping density is N_d. The surface of the semiconductor is at $x = 0$ and a depletion region is formed extending to a distance W within the semiconductor. As the semiconductor is depleted of mobile carriers in the depletion region, the charge density in that region is due to the ionized donors (ID) or ionized acceptors (IA). Hence

$$\rho(x) = eN_d \quad 0 < x < W$$
$$\rho(x) = 0 \quad x > W. \tag{1a}$$

Solving Poisson's equation

$$\nabla^2 V(x) = \rho(x)/\varepsilon_s \tag{1b}$$

gives

$$V(x) = \begin{cases} \dfrac{eN_d}{2\varepsilon_s}\,[W^2 - (W-x)^2] & 0 < x < W \\ \dfrac{eN_d}{2\varepsilon_s} W^2 & x \geq W, \end{cases} \tag{2}$$

where $\rho(x)$ is the charge density in the space charge region, of width W and x is the distance from the interface of interest (hetero- or homojunction, metal/semiconductor interface, or charged surface) and ε_s is the dielectric constant of the material.

From Eq. (2) we can calculate the depletion layer width W, as

$$W = \sqrt{(2\varepsilon_s V_{\mathrm{bb}}/eN_{\mathrm{d}})}. \tag{3}$$

Here, eV_{bb} is the band bending corresponding to the value of $V(x)$ at $x = W$. Note that the solution of Poisson's equation for our physical situation produces a quadratic dependence of the energy with depth as mentioned above. This means that the bands within the semiconductor are "bent" parabolically. As we will describe, the band bending (eV_{bb}) can be measured in an FPS experiment. In addition to determining the band bending we can see from Eq. (3) that when the doping concentration is known, our knowledge of the magnitude (and direction of the band bending) indicates that we can also know the depletion width.

6.4 Band Bending at the Metal–Semiconductor Interface

In the case of a metal–semiconductor Schottky barrier, charge transfer is induced through a mismatch in the work functions of the metal and semiconductor. A typical case is where a relatively high work function metal accepts electrons from an n-type semiconductor whose work function is smaller. In this case, the semiconductor will be charged positively and the metal negatively. This results in upward bending of the semiconductor bands; the

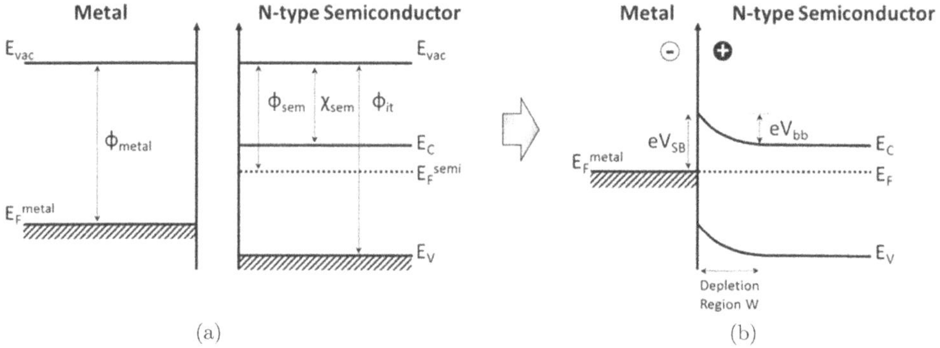

Fig. 2. (a) Schematic of separated large work function metal and n-type semiconductor. Since they are separated they share a common vacuum level (E_{vac}). (b) upon contact between the metal and semiconductor electrons flow from the semiconductor to the metal, equilibrating the Fermi levels resulting in a negatively charged metal and upward bending.

situation is shown in Fig. 2. Figure 2(a) shows the separated metal and semiconductor. The two materials share a common vacuum level (E_{vac}). X_{sem} is the electron affinity of the semiconductor and Φ_{it} is the ionization threshold.

When the two materials are coupled, electrons flow from the n-type semiconductor (in this case) to the metal until their Fermi levels equilibrate. A consequence is that a barrier to electron flow from the metal into the semiconductor is formed, known as a Schottky barrier (eV_{SB}).

Heterojunctions are the more generalized term when discussing the couple formed between two dissimilar systems. A simple version as discussed above is that of a Schottky junction consisting of a metal deposited on a semiconductor. Typically, the work function of the metal and that of the semiconductor are different. In Fig. 2, Φ_{sem} is the work function of the semiconductor, Φ_{metal} is the work function of the metal, and Φ_{it} is the ionization threshold of the semiconductor.

We delve more deeply into the case where the metal has a smaller work function than that of, for example, a p-type semiconductor. As before when the metal and semiconductor are infinitely separated, they share the same vacuum level but their Fermi levels are different. There is no electrical connection between the two materials, so they have no "knowledge" of the location of each other's Fermi level. Recall that the work function of a material is the energy it takes to promote an electron at the Fermi level to the vacuum level with zero kinetic energy. As we bring the two materials together from infinity and couple them, physics tells us that the Fermi levels must equilibrate. To do so, electrons from the lower work function metal will transfer into the higher work function semiconductor. At equilibrium, the Fermi levels are the same but the metal is left positively charged while the semiconductor is negatively charged. In an equivalent argument relative to our earlier discussion, the bands will bend downward. In the relatively uncomplicated case where no states are formed at the interface between these two materials (Schottky limit), we can simply calculate the band bending from the charge transferred from the metal to the semiconductor. But, of course life is not so simple and the pure Schottky picture rarely holds. This is because in reality, defect states do form at the metal–semiconductor interface resulting in additional trapped charge that, if sufficiently numerous, can "pin" the Fermi level within the band gap at the surface resulting in a deviation from the basic Schottky picture.

6.5 Fermi Level Location in Semiconductors

When the materials are not electrically connected, their electronic energy levels can only be referenced to the vacuum level (E_{vac}), the energy level at which a liberated electron exists at rest outside of the material (for example, through the absorption of a photon). The work function of the metal, Φ_{metal} is the minimum energy required to promote an electron from the Fermi level to a state outside the material with zero kinetic energy. For a semiconductor, this case is a bit more complicated. The Fermi level of a semiconductor is dependent upon the concentration of ionized acceptors, IA or ionized donors, ID. As a result, the work function of a semiconductor is then dependent upon its Fermi level location that is in turn dependent upon the doping level. Hence, from this argument a *p*-type semiconductor will have a larger work function than an *n*-type semiconductor. When UPS is carried out on a semiconductor the onset of electron emission, or ionization threshold (it) is given by Φ_{it}. If we plot this threshold relative to the Fermi level we will see that it shifts depending upon the nature of the doping (*p*- or *n*-type) and the concentration of dopants. This effect is shown in Fig. 3.

The location of the Fermi level can be calculated based upon intrinsic semiconductor properties and the statistics of doping. A thorough treatment is given in Sze [5] and we capture the specific details salient to our discussion below.

To start, the Fermi level for an intrinsic (undoped) semiconductor lies close to the center of the band gap. Introduction of dopants results in a shift of the Fermi level in order to maintain charge neutrality. Typically, donor (acceptor) atoms are chosen so that they form shallow energy levels relative to the conduction (valence) band edges so that at temperature T (typically

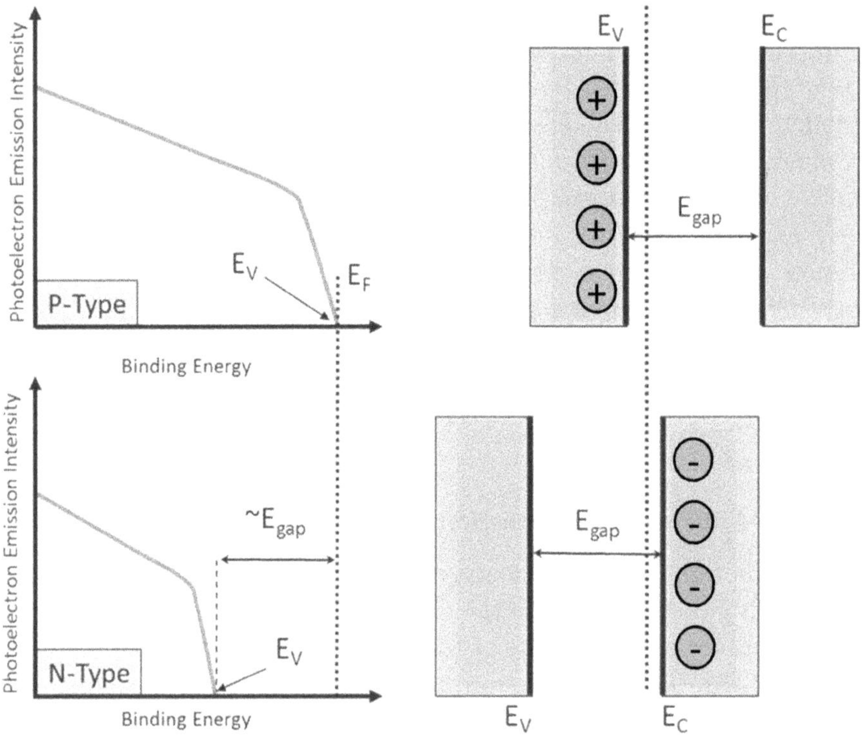

Fig. 3. (top) UPS schematic spectrum of a p-type semiconductor. The Fermi level is found close to the valence band maximum due to an excess of holes within the valence band. The schematic electronic structure is shown to its right. (bottom) UPS schematic spectrum of an n-type semiconductor. The Fermi level is now located close to the conduction band minimum due to an excess of electrons within the conduction band.

room temperature of 300 K), most are ionized. The number of ID or IA in the conduction (valence) band is given by

$$N_{ID}^{+} = N_D \left[1 - \frac{1}{1 + \frac{1}{g} \exp(\frac{E_D - E_F}{kT})} \right] \tag{4}$$

$$N_{IA}^{-} = \frac{N_A}{1 + g \exp\left(\frac{E_A - E_F}{kT}\right)} \tag{5}$$

where N_{ID}^{+} is the number of ID, N_D is the density of donor dopants and similarly, N_{IA}^{-} is the number of IA while N_A is the density of acceptor dopants. The factor g is the ground state degeneracy and is dependent upon details of the semiconductor bands. The key issue here is that for shallow donor or acceptor levels the degree of ionization approximates the density of dopant atoms introduced into the semiconductor. Finally, from requiring neutrality we can write

$$N_C \exp[-(E_C - E_f)/kT] = N_V \exp\left[(E_V - E_f)/kT\right] + N_D \left[\frac{1}{1 + 2\exp\left(\frac{E_F - E_D}{kT}\right)}\right], \qquad (6)$$

where N_C and N_V are the effective conduction and valence band density of states given by the effective mass model. A similar calculation can be done for acceptors. N_C and N_V are tabulated for most semiconductors and for more exotic materials, as we will describe when we discuss multinary photovoltaic absorber materials, these values can be extracted from *ab initio* calculations.

From Eq. (6) the Fermi level can be calculated and compared with, for example, a UPS experiment. Typically, UPS is not often used because the bands near the surface may be bent in response to surface charge. As described previously, a typical semiconductor has charged defect states that bend the bands near the surface. For standard UPS, this complication interferes with the accurate location of the Fermi level within the semiconductor because band bending will shift states near the surface closer to or further from the Fermi level.

An accurate location of the Fermi level can only be obtained when the surface is "flat-band", i.e. when the band bending is removed. As we will describe, flat bands can be obtained by photoexcitation of the semiconductor with a laser pulse of sufficient intensity to completely screen out the existing static dipole. This process, described later is the key focus of this chapter.

Another potential complication to accurately determining the Fermi level with photoelectron spectroscopy is that the atomic structure at the surface can differ from that of the bulk. Occupied surface states on elemental semiconductors as well as compositional differences in multinary (multi-elemental) semiconductors are complications; almost all semiconductor surfaces are reconstructed. This effect can to some degree be ameliorated by the methods used for extracting the actual valence edge. GaAs was one of the first [6] semiconductors used to extract the location of the valence band maximum by comparing the photoelectron spectrum with *ab initio* calculations of its electronic structure. Later [7] studies showed that a simple extrapolation of the valence edge by a least squares fit and later a simple linear fit gave accurate results of the location of the valence band maximum. The linear fit approximation is used most extensively in this chapter.

Interferences associated with non-ideal surface layers can be handled by considering the nature of the layer. If the non-ideal layer is very thick, i.e. much greater than the electron escape depth then it is likely impossible to identify the valence edge of the underlying semiconductor. But the typical case involves a layer that is 1–2 atomic layers thick, such as atomic layer reconstruction, or a thin oxide layer. For atomic layer reconstruction, the underlying semiconductor valence structure can still be observed and extrapolation can be carried out. For oxides, typically the valence edge of the oxide lies at a higher binding energy and interferes with the underlying semiconductor valence edge only by reducing its intensity. Adjusting the exciting photon energy can be used to increase the electron escape depth. Referring to Chapter 4, low and high photon energies can provide an escape depth of 1–2 nm and underlying valence structure can be extracted for several escape depths.

Another possibility involves passivation of the surface to remove oxides and surface reconstruction. A classic example for Si surfaces involves the etching of Si in dilute HF (typically 10% HF in distilled H_2O) that removes native oxide and hydrogen terminates the surface. The H termination also removes any surface states that might interfere with the determination of the valence band maximum.

6.6 Femtosecond Photovoltage Basics

Several examples that describe the basics of FPS are described in the following sections. The process of determining the electronic structure and, in particular, the lineups of band edges in multilayer systems involves, as will be explained in greater detail, a set of sequential measurements. The steps are listed below:

1. Determine valence band maximum of the substrate semiconductor under flat band conditions. This is achieved by fs-UPS of the clean (unoxidized) substrate.
2. Determine the valence band maximum of the overlayer semiconductor or oxide under flat band conditions. This is achieved by fs-UPS where the pump pulse is transmitted through the top layer and absorbed by the semiconductor substrate.
3. Comparison of the unpumped and pumped spectra determines the band bending of the substrate semiconductor.
4. Accurate determination of the Fermi level of the system by UPS from a clean metal in electrical contact with the semiconductor system or in the case of a metal/semiconductor system, direct measurement of the top metal.

The case studies below provide direct examples of the process used to determine the electronic structure of complex multilayer systems. In contrast with the standard approach that involves the monitoring of atomic core levels following sequential depositions of increasingly thick layers of material atop a semiconductor substrate, the approach described below can be carried out with a single measurement of *ex situ* fabricated structures.

6.6.1 *Semiconductor Heterostructures*

Consider a basic heterostructure system in which two dissimilar materials are coupled. The bottom semiconductor, for the purposes of this example, is p-type but our discussion is equivalent for an n-type semiconductor as well. The top layer could be a metal whose work function is smaller than that of the semiconductor and this would result in the formation of a Schottky contact. In the more general case, as shown in Fig. 4 the top case is an n-type semiconductor, forming a p–n junction that is the basis for bipolar transistors, photovoltaics, and other electronic devices. During the formation of the p–n junction, electrons flow from the n-type top semiconductor to the p-type substrate to equalize the Fermi levels of the two materials and in doing so leaves the top layer positively charged relative to the substrate. This charge transfer results in the formation of a dipole field within the semiconductor that bends the bands downward (p-type) or upward (n-type) as described earlier in this chapter.

We wish then to interrogate this system to determine the location of the Fermi level, the locations of the valence band maxima for the substrate and top layer semiconductors and the magnitude and direction of the band bending to fully characterize the electronic structure of the system. This is shown in the sequence of schematic electronic structure in Fig. 4.

In Fig. 4(a), an XUV pulse of light impinges on the semiconductor stack and produces a UPS spectrum of the top semiconductor. The Fermi level of the system is determined through measurement of a clean metal in electrical connection to the sample, allowing the UPS spectrum to be displayed on a binding energy scale with the Fermi level set at 0 eV.

In Fig 4(b), a synchronized pump pulse passes through the wider gap top semiconductor and is absorbed in the underlying substrate semiconductor. Absorption of this pulse excites a dense electron-hole (e-h) plasma and the carriers respond to the existing internal dipole field in such a manner that they produce an opposing field that flattens the bands of the substrate semiconductor. Electrostatics shifts the energy bands of the top layer as well and we observe this shift in the UPS spectrum originating from the top

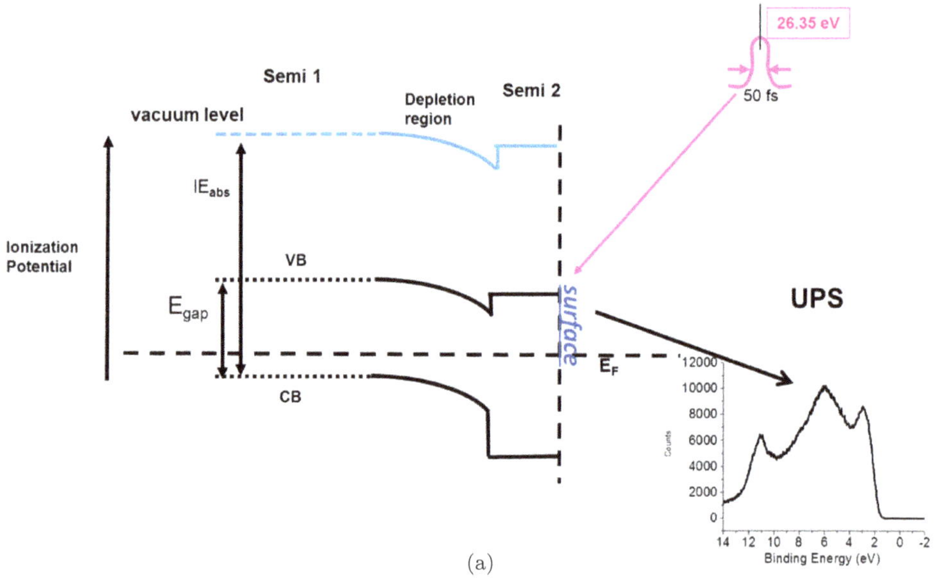

(a)

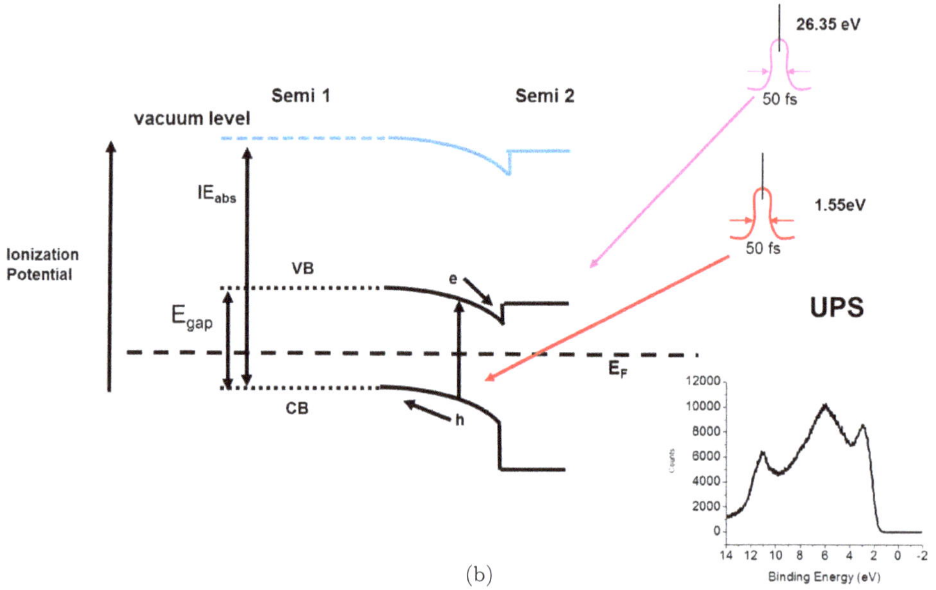

(b)

Fig. 4. Semiconductor heterostructure system showing the stepwise (a)–(d) process of photoexcitation (pump) and band flattening using synchronized pump and XUV probe pulses. (a) XUV spectrum of top semiconductor. (b) Excitation pulse generates dense e-h plasma. (c) Substrate semiconductor bands flatten, rigid shift of spectrum (red) relative to probe-only spectrum (black) provides band bending magnitude and direction. (d) Direct comparison of flat band substrate and top semiconductors revealing band edges.

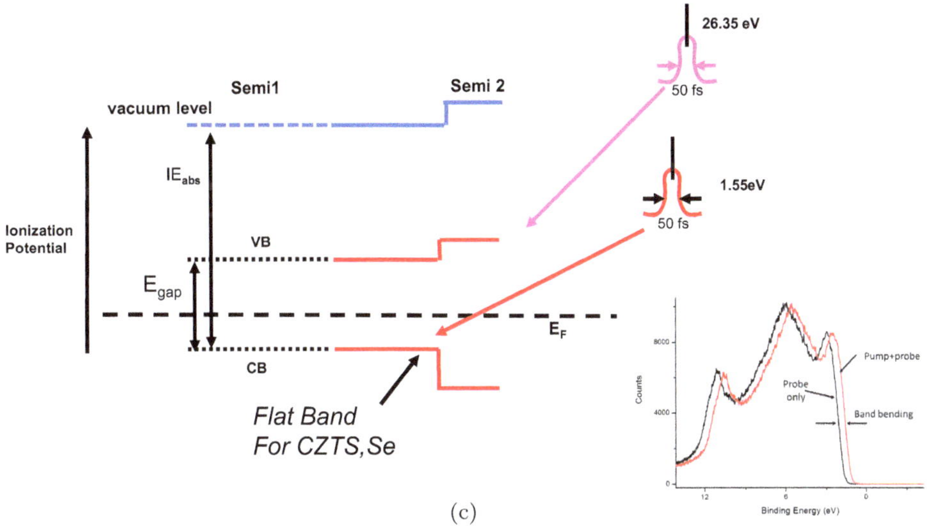

(c)

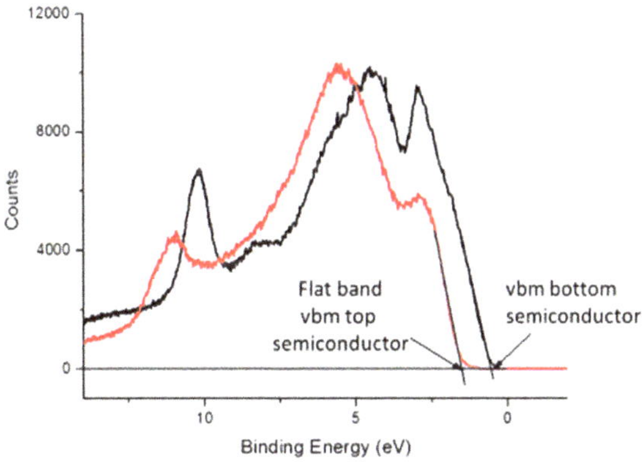

(d)

Fig. 4. (*Continued*)

layer. This is shown schematically in Fig. 4(c). The red spectrum is shifted by the amount that the bands flatten. Finally, in Fig. 4(d) we compare the flat band valence band maxima of the substrate and top layer semiconductors; with this information the band lineups and Fermi level location of the heterostructure is determined.

Utilizing intense pulses from the amplified femtosecond laser drives this band flattening to saturation. As such, the shift of the photoexcited spectrum relative to the unexcited spectrum provides both the magnitude and direction of the band bending. This effect is self-limiting, i.e. it is not possible to over-excite the system because once the internal depletion field is eliminated by the presence of the excited e-h plasma any additional electrons and holes will be created in a field-free environment. A remarkable aspect of this process is that even though UPS is a highly surface sensitive technique and we are flattening the bands of the underlying "buried" semiconductor, we can observe this same energetic shift in the top layer.

6.6.2 *Metal-Oxide-Semiconductor Stacks*

Here, we consider a somewhat more complex system — that of the metal-oxide-semiconductor or MOS stack. The layout of an MOS, field-effect transistor (MOS-FET) is shown in Fig. 5 [5]. The figure depicts a metal deposited on a thin gate oxide (SiO_2) residing on n-type Si. At the gate oxide/Si interface the Si bands will be bent upward in response to negative charge within the oxide, modified by electron transfer between the semiconductor and gate metal. This structure is called a P-FET.

A key question for technologists hoping to make a low threshold voltage MOS-FET device from this structure is: Where is the Fermi level located at the Si/oxide interface? The location of this interface Fermi level (IFL) is critical since the closer the top of the valence band level is to the Fermi level near the gate dielectric interface, the lower the applied gate voltage needed to initiate the flow of holes. In Fig. 6, we show the electronic structure

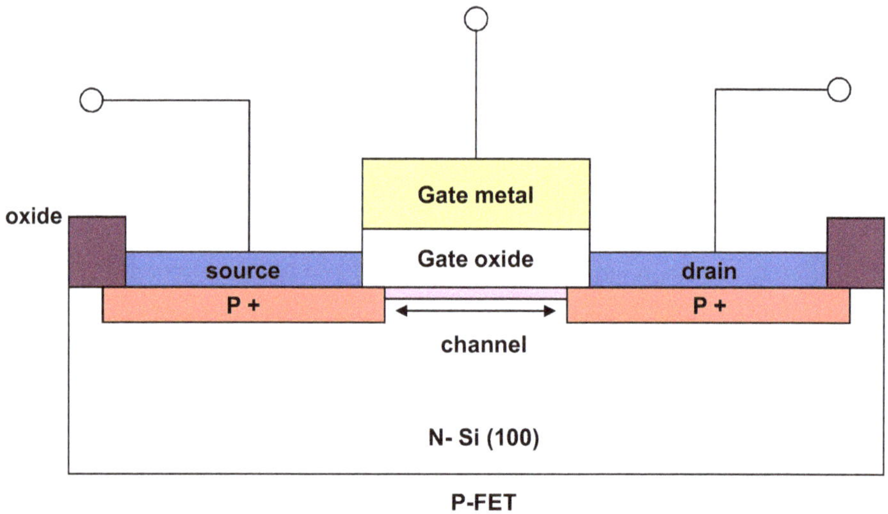

Fig. 5. Schematic of a p-FET (p-channel field effect transistor). A gate voltage will draw the Si valence bands up, opening a channel for the flow of holes from source to drain.

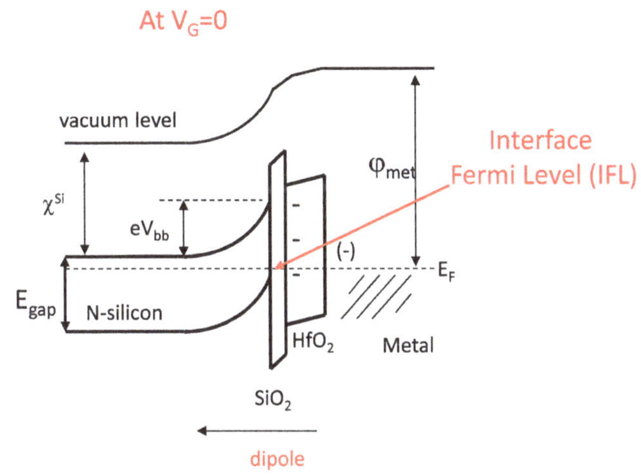

Fig. 6. Energy level schematic of a Si MOS structure. Electron transfer from the n-Si to the metal forms a dipole that bends the Si bands upward. Additional charge within the oxide, shown as a combination of a thin SiO_2 layer and a thicker HfO_2 layer contributes to the field.

of the Si/oxide interface depicting the upward bent bands. In this case, the channel is formed when a negative bias is applied to the gate metal pulling the bands upward. As the valence band maximum approaches the Fermi level holes fill the channel. The holes in the channel are transported perpendicular to the plane of the page in Fig. 6 between the source and drain shown.

This fundamental device structure has been studied in extensive detail over many years [8]. Capacitance-voltage (CV) [5, 9–13] measurements carried out on fabricated capacitor structures are the gold standard for determining the IFL and capacitance. One drawback is that actual capacitive structures must be fabricated in a clean-room fab that can take weeks to produce, but is one of the few methods for extracting the IFL because accessing the Fermi level location is difficult as it is buried under metal and oxide layers. In addition, modeling methods are usually required to identify the location of the Fermi level at the interface.

It is here that FPS spectroscopy can solve a critical technology measurement problem. First, consider Fig. 7 where energetic light, typically above 10–30 eV is directed to the surface of the material stack. This is standard UPS and the spectrum generated, as shown in Fig. 7(b) is that of the metal gate material, in this case Pt [13, 14].

While UPS is used to establish the Fermi level of the system by generating a spectrum from the gate metal, we have no knowledge of the properties

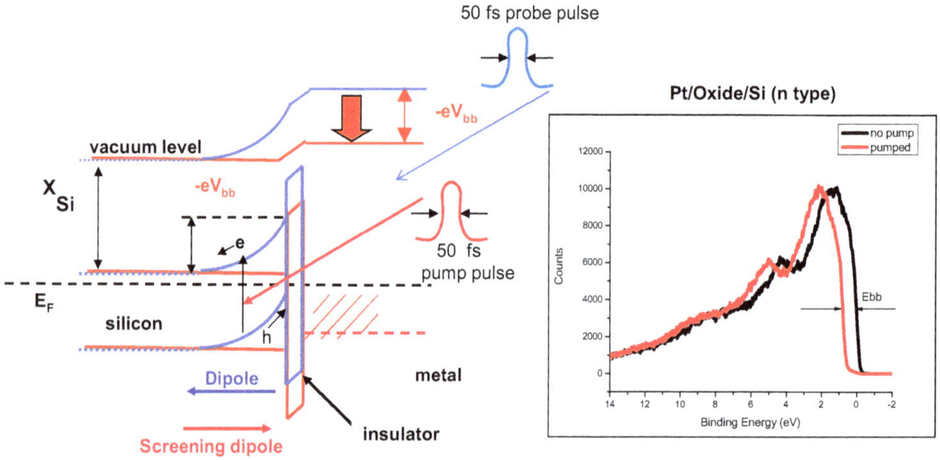

(a) (b)

Fig. 7. (a) Energy schematic showing the semiconductor bent bands, gate dielectric, and metal. Pumping with a femtosecond pulse of light flattens the bands (shown in red). (b) Unpumped and pumped UPS spectra. The rigid shift of the spectrum upon pumping reveals the magnitude and direction of the band bending.

of the underlying gate insulator or semiconductor. In order to determine the properties of the entire system we need measurements of the individual parts under flat band conditions. This would involve, for example, knowledge of where the Fermi level is within the semiconductor and what effect the presence of the insulator has on band bending within the semiconductor as well as the metal when the final gate stack is completed.

As we discussed earlier in this chapter, the location of the Fermi level within the semiconductor can be ascertained through knowledge of the doping level. For Si and other heavily utilized semiconductors like Ge and III–V materials such as GaAs, InP and more, doping levels and Fermi level locations within the semiconductor are well tabulated [5] and can be found in archives, such as the Ioffe Physico-Technical Institute Electronic Archive (http://www.ioffe.ru/SVA/NSM/). In addition, also described above, the Fermi level can be obtained with fs-UPS wherein the clean surface is pumped to achieve a flat band condition and the location of the valence band maximum is determined relative to the Fermi level [7]. An example of this is shown in Fig. 14 for n- and p-doped Si.

FPS is carried out by excitation with a visible pulse of femtosecond light time correlated with the XUV pulse for photoemission. As described in Chapter 2, the output of a pulsed femtosecond laser system, typically

Ti:sapphire produces 800 nm (1.55 eV) light whose energy is sufficient to pump Si and most III–V semiconductors. For higher band gap materials such as GaP, CdTe, CdS and more, the 800 nm light can be frequency doubled to 400 nm (3.1 eV). With sufficient intensity, the otherwise bent bands of the semiconductor, either clean or covered with a dielectric (typically transparent due to its wide band gap) and thin gate metal are transiently flattened. For our experiments, the gate metal is kept deliberately thin so that sufficient pump light intensity reaches the underlying semiconductor.

Synchronized fs-XUV pulses, produced via high-harmonic generation are used to photoemit electrons from the material. Since the gate metal is thicker than the mean-free path of the photoelectrons, we only see the metal states and can clearly identify the Fermi level which we assign to be zero energy. This is typically done in photoelectron spectroscopy since, as described in Chapter 4, both the sample and detector share the same Fermi level; hence, the Fermi level is the one constant and "anchor point" of the experiment. Since the Fermi level is given the arbitrary value of 0 eV, the photoelectron spectrum of occupied valence states is plotted as a function of binding energy.

Figure 7 captures the essential physics of the process and relates to Fig. 4. The static structure is shown in blue with bands bent upward for the n-type semiconductor. Electrostatics dictates the energetic properties of the system as shown.

Upon pumping, the photogenerated electrons and holes immediately respond to the existing dipole field responsible for the bending of the semiconductor bands. The electrons are driven into the semiconductor interior while the photogenerated holes are driven toward the surface. For a p-type semiconductor this process would be reversed. As the electrons and holes redistribute, they produce a dipole field that opposes the original static field; the sum of the two results in a field-free condition and the bands flatten. As can be seen from Fig. 7, the original energy schematic shifts downward in energy upon pumping (red spectrum and schematic). Since the effect is purely electrostatic in nature, driven by the eradication of the dipole field, the entire photoelectron spectrum shifts in response. This rigid shift is seen in the comparison of the pre- (blue) and post-excitation (red) spectra. The downward direction of the shift tells us that the bands of the underlying semiconductor are originally bent upward typical of an n-type material. Hence, both the magnitude of the band bending and semiconductor type are revealed in this measurement. There are exceptions to this case as we will discuss later but in general the magnitude and direction of the shift can be simply read from the comparison of the spectra. Hence the underlying semiconductor band bending can be fully characterized even though the semiconductor is buried beneath layers of metal and oxide!

6.7 Detailed Considerations: How Femtosecond Photovoltage Spectroscopy Works

Now we consider the details of fs-UPS. We first approximate the sample-detector system as a parallel plate capacitor. A generalized case is, as above, an MOS stack. The potential of the top metal layer is constant before photoexcitation, and at a constant potential difference with respect to the potential of the detector as was discussed in Chapter 4. There exists a static electric field between the sample and the detector corresponding to this potential difference. After photoexcitation, the potential of the metal at the excitation spot changes by the amount of the induced photovoltage, and correspondingly the electric field between them. Photoemitted electrons are accelerated (decelerated) in the modified electric field, and gain (or lose) kinetic energy by the amount of the photovoltage. As a result, we can extract the photovoltage by measuring a probe-only spectrum and comparing it to the energetically shifted spectrum collected from the photoexcited (pumped) sample. If there is a change in the potential of the sample due to carrier relaxation during the time the electrons are transiting to the detector, the already emitted electrons near the sample will be influenced by the change in the electric field. This time dependence is discussed in the next section, but is inconsequential for our femtosecond timescale here. Since the excitation spot is much smaller than the entire sample, the change in the electric field due to the photovoltage is localized near the excitation spot. The range of this electrostatic effect is therefore limited and depends on geometrical factors, primarily the size of the pump spot.

To analyze the problem more quantitatively, we numerically solved Poisson's equation using the relaxation method for the boundary problem to calculate the potential distribution near the sample [4]. For simplicity, we treat the sample and detector as two parallel electrodes of a capacitor, both grounded. The potential is constant inside the capacitor and there is no electric field before photoexcitation.

The potential distribution after photoexcitation, calculated for an excitation spot diameter of 2 mm and a photovoltage of 1 V, is plotted in Fig. 8(a).

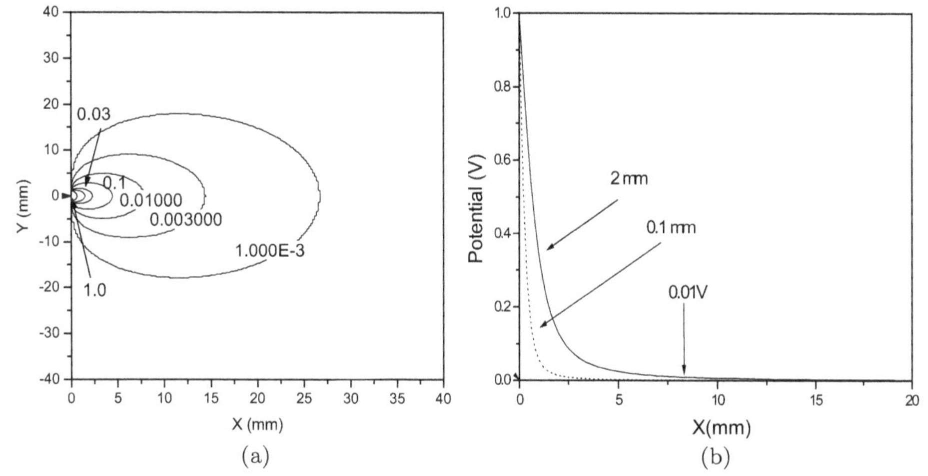

Fig. 8. (a) Potential distribution between the sample and the detector, which was simplified as a parallel plate capacitor. Sample plane is at $x = 0$, and the diameter of the excitation spot centered at (0, 0) is 2 mm. The potential of sample and the detector electrode is set to 0 except for the excitation spot, which was set to 1 V. (b) The potential profiles are plotted as a function of the distance X from the sample, for excitation spot sizes 2 mm and 100 μm.

The potentials along the center line between the sample and detector are plotted for excitation spot sizes of 2 mm and 100 μm in Fig. 8(b). The potential drops rapidly near the sample and almost vanishes at distances farther than the excitation spot diameter. It will be convenient to assign a characteristic time (length) scale: the electron escape time (length) as the time (length) required for the electron to travel to a position, where the potential drops to 1% of that at the excitation spot. If the photovoltage suddenly disappears when the electron is at this position, it will cause a maximum 1% error in the photovoltage measurement. For our given experimental geometry, the electron escape length is ~8 mm. The escape time, on the other hand, depends on the kinetic energy of the electron. In typical fs-UPS experiments, photoelectron spectra were collected with 26.35 eV photons and electrons emitted from states near the Fermi level with kinetic energies of ~22 eV were used to extract the photovoltage shift. For these electrons, the escape time to 8 mm is ~2.5 ns. Any changes in the sample on a timescale longer than 2.5 ns would have a negligible effect on the measured electron energies. Since the e-h recombination time in crystalline Si is $\geq 10^3$ times longer, we expect no impact on the photovoltage measurements described here. But as we have mentioned, even semiconductor systems with ns lifetimes, uncertainties of just a few percent in the actual band bending are within the uncertainties (~10–30 meV) in extracting the band bending from the UPS spectra.

We can reduce the escape time by using higher photon energies (and hence higher electron kinetic energies) generated with more energetic X-rays, or by reducing the size of the photoexcitation spot. For example, if we increase the probe beam to 5 keV, the escape time will be only ~170 ps. On the other hand, if we decrease the excitation spot to 100 μm, the electron escape time (length) will be ~600 ps (2 mm).

For accurate photovoltage measurement, the photovoltage decay must be slower than the electron escape time. Several pathways of photovoltage decay can be envisioned: ultrashort photoexcited carrier lifetime and the screening time of the potential change in the metal. The density of the non-equilibrium carriers, responsible for the photovoltage, decreases via several pathways, such as e-h recombination, Auger recombination, defect trapping, and surface recombination. For photovoltage measurements, the lifetime should be longer than the electron escape time. For semiconductors with short lifetimes, such as low temperature grown GaAs, the measurement error can be significant. For example, for a sample with a lifetime of ~100 ps, the measured photovoltage shift can be underestimated by as much as 30% of the actual value.

An additional consideration is that of screening effects in the metal. After photoexcitation, the potential of the metal in the illuminated spot will be decreased (increased) by the amount of photovoltage, and electron transport into (out of) the photoexcited spot occurs until the potential gradient disappears in the metal film. This screening time depends greatly on geometric factors since electrons will move in from the edge of the excited spot. The gradient of the potential exists only at the boundary, and thus the field forcing electrons to move into (out of) the spot exists only at the boundary. If we make a simple assumption that the field is constant across the excitation spot, we expect a screening time of ~ 1 μs for a photovoltage of 1 V and a 2 mm excitation spot, and hence can be ignored.

6.7.1 *Pump-Probe Delay Dependence*

Photovoltage shifts were measured for several pump-probe delays as shown in Fig. 9. The photovoltage shifts are almost constant from −30 ps to +160 ps as expected. This confirms the fact that we measure the time integrated average of the photovoltage, rather than the instantaneous photovoltage at the delay time. The relative independence of the photovoltage on the pump-probe

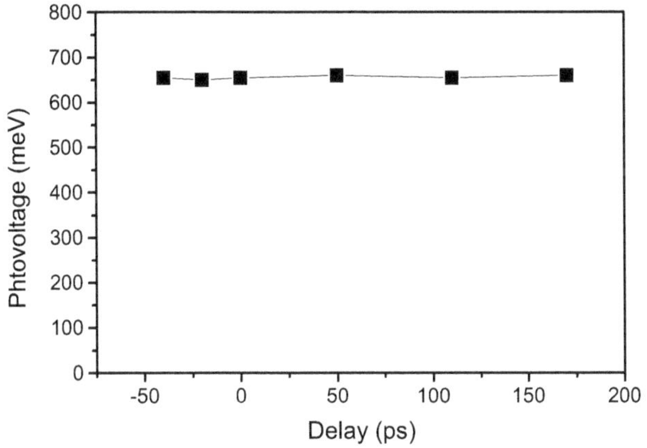

Fig. 9. Change in photovoltage shift as a function of pump-probe delay for Si. The constancy of the shift reflects the relatively long lived e-h populations following photoexcitation.

delay also confirms that the carrier lifetime in silicon is longer than the electron escape time for our experiment.

6.7.2 *Doping Density Considerations and Saturation of the Photovoltage Shift*

Several interesting and important considerations of this band flattening are worth noting. First, there is a self-limiting aspect of this process. Once the static dipole field is eradicated, additional excitation of electrons and holes occurs in a field-free condition, hence the effect is asymptotically self-limiting. As a result, it is not possible to "pump" the system too hard, aside from other deleterious effects at high laser excitation densities such as heating or melting. A second consideration is that of the background doping level of the semiconductor. Typical pump photoexcitation e-h densities can reach up to and beyond $10^{20}/\text{cm}^3$. Above this density, heating and eventual melting of the material becomes a concern. At or below this level, calculations have shown that the sample in the region of the excitation spot heats by only a few degrees K and so is not a concern.

Even at this excitation density, if the background doping density is $> 10^{18}/\text{cm}^3$ the bands may not fully flatten and corrections must be made to ascertain the actual band bending. To treat this effect, we describe the following treatment of band flattening.

If the bands are completely flattened by photoexcitation, the band bending is the same as the measured photovoltage shift. For lightly doped semiconductors and strong photoexcitation, this condition holds and no correction is necessary to extract the band bending. However, this is not the case when the photoexcited e-h density is comparable to the doping level.

Consider an experiment in which the actual band bending is Y_o but with insufficient pumping or too high a doping level we only partially flatten the bands, leaving a residual band bending of Y [4]. This situation is displayed in Fig. 10. The actual spectral shift we would observe, then is $Y_o - Y = \Delta V$. Once we calculate Y, the residual band bending, we can add this value to ΔV to obtain the actual band bending.

The original treatment of this problem was published by Johnson [15] and later by Long [16]. Here, we provide an approximate solution to the residual band bending in the absence of changes in the surface charge density during the measurement. Since we are utilizing femtosecond pulses and measuring

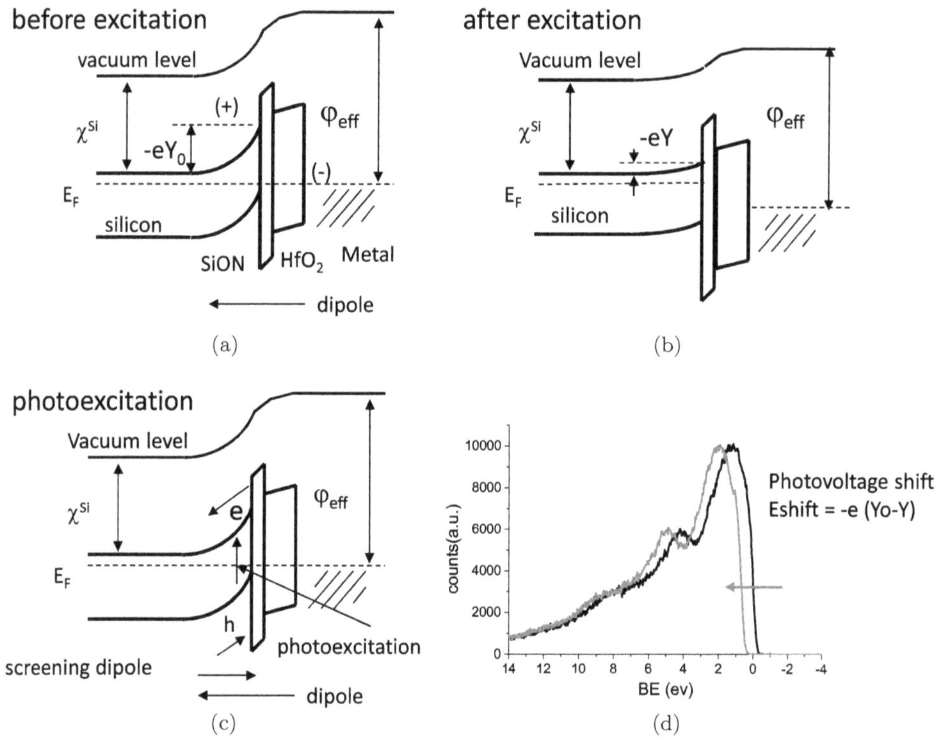

Fig. 10. (a) MOS stack in equilibrium. eY_o is the band bending in the Si substrate. (b) Initial photoexcitation–photoexcited electrons and holes react to the internal field. (c) following photoexcitation, bands flatten — if excitation density is insufficient to fully flatten bands, residual band bending is given by eY. (d) UPS from metal for probe only and pumped showing the rigid spectral shift as the underlying Si bands flatten [4].

near $t = 0$, this is a valid approximation. For the case where the actual band bending, $eY_o >> \mathrm{kT}$ (26 meV at room temperature), the residual band bending following photoexcitation, $eY > \mathrm{kT}$, the doping density in the semiconductor is N_d, and the excited density of electrons and holes is N_{exc} we can express the residual band bending as

$$eY = \mathrm{kT}\ln[\Delta V N_d/\mathrm{kTN_{exc}}] \tag{7}$$

The results of this calculation are shown in Fig. 11. As can be seen Eq. (7) provides an accurate value of the actual band bending in the semiconductor.

Excitation densities possible from an amplified femtosecond laser source can be quite high-certainly in excess of $10^{20}/\mathrm{cm}^3$. Nonetheless, for very high doping densities some errors due to incomplete band flattening may occur. The impact of the above discussion is that for semiconductor doping levels $< 10^{17}/\mathrm{cm}^3$ and high photoexcitation densities which can reach $\leq 10^{20}/\mathrm{cm}^3$ without appreciably heating the sample, the band bending in the semiconductor can be taken directly from the rigid shift in the pumped photoelectron spectrum relative to the unpumped with high accuracy. For doping densities higher than $\sim 10^{17}/\mathrm{cm}^3$, Eq. (7) can be applied to correct for the small residual field to yield an accurate semiconductor band bending.

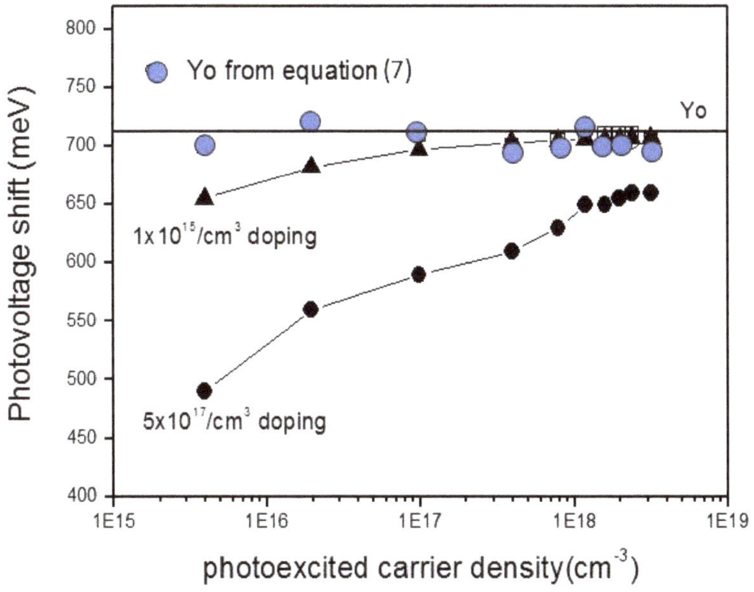

Fig. 11. Comparison of photovoltage shifts as a function of photoexcited carrier density for two different doped semiconductors. Using Eq. (7) the actual band bending (720 meV) can be determined from the measured photovoltage shifts [4].

As a final consideration, we discuss *saturation* of the photovoltage shift. Since we are using an intense pulsed laser source and not a lamp, exceedingly high excitation densities can be achieved. As mentioned previously, and in particular for femtosecond laser sources, excitation densities are typically limited to $\sim 10^{20}/\text{cm}^3$ or below; above this level the temperature of the sample will rise and eventually ablation (sample damage and material removal) would occur.

Experimentally, it is important to work within the "saturation" region of the excitation intensity. As described above, there is an electron "escape depth" external to the sample given by the recombination of the excited e-h population in the sample which can be several mm outside of the sample itself. However, this escape depth is dependent upon the actual e–h density at any given time. If we excite, for example, the sample to a density of $5 \times 10^{19}/\text{cm}^3$ e–h pairs at $t = 0$ (the time of photoexcitation) it may take many nanoseconds for the excitation density to drop below a level where the internal dipole field that bends the bands will reform. Above this level, the bands will remain flat. The specific conditions for this would have to be calculated for every semiconductor.

A simpler method is to simply measure the change of the photovoltage shift with increasing excitation density. This is shown in Fig. 12. As can be seen, the band bending shift increases with increasing pump fluence until a level of excitation is reached where the bands shift no further. The saturation

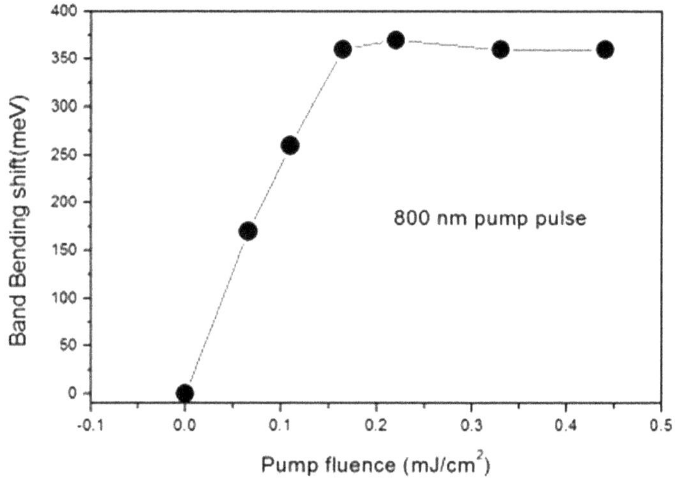

Fig. 12. Saturation of band bending shift as a function of pump fluence for Si. Beyond $\sim 0.2\,\text{mJ/cm}^2$ no further spectral shift is observed indicating that the bands are fully flattened. A simple series of measurements such as this confirm the pump fluence required to fully flatten the semiconductor bands [4].

occurs because, as described above, the bands are flat and the internal dipole field is completely screened. Additional excitation occurs within a field-free condition where the electrons and holes are no long driven apart. Experimentally, operating within this saturation region will result in the most accurate extraction of the band bending magnitude in the sample under study. Given that the collection of each spectrum for a given excitation density takes only a few minutes, such a graph can be generated in a very short period of time.

6.8 Measurement of Work Function with UPS

One additional measurement that can be important to fully characterize the electronic structure of a system is the work function of the top-most material. The work function is determined from knowledge of the photoemitting photon energy, the location of the Fermi level and the width of the photoelectron spectrum; this can be expressed through the equation

$$\Phi = \hbar\omega - W_{\text{spectrum}}, \tag{8}$$

where W_{spectrum} is the width of the photoelectron spectrum and $\hbar\omega$ is the photoemitting photon energy. Figure 13 shows a schematic of this.

Electrons emitted at the vacuum level have zero kinetic energy and as a result are not collected by the electron spectrometer; this problem is solved by biasing the sample negatively with respect to the spectrometer which shifts the entire spectrum to higher kinetic energy. Determining the work function of the top material accurately requires that the surface is free from

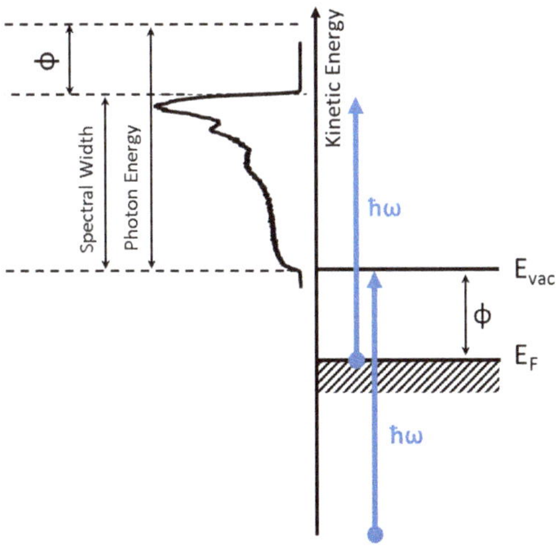

Fig. 13. Schematic showing the measurement of the work function of a material with UPS.

oxides or other contaminants but with this approach Φ can be determined with great accuracy.

6.9 Summary of Femtosecond UPS Photovoltage Measurements

Before continuing on to case studies we summarize what has been discussed. Using intense femtosecond pulses of light to photoexcite a semiconductor where the surface is clean, oxide or metal+oxide covered, the bands can be fully flattened; this is observed as a rigid energetic shift of the pumped spectrum relative to the unpumped. The direction of the shift reveals whether the semiconductor bands were originally bent downward (p-type) or upward (n-type). As long as the density of photoexcited electrons and holes is much greater than the background doping density the extraction of the band bending is accurate. For particularly high doping densities a simple correction (see Eq. (7)) can be applied to correct for small residual fields still existent in the semiconductor, but well above $10^{18}/\text{cm}^3$ doping densities determination of the actual band bending becomes increasingly inaccurate. Semiconductor band bending can be accurately extracted even under thin layers of oxide and metal. For oxides above 5–10 nm, accuracy is reduced due to possible charging of the oxide itself. For metals deposited on top of the oxide, thicknesses above 10 nm can result in significant reduction in the light intensity reaching the semiconductor and this can impact the degree of band flattening and hence render the measurement inaccurate. Fermi level location within the band gap of a semiconductor can be accurately determined under flat band conditions leading to a method for determining the Fermi level in typical and non-standard (e.g. multinary) semiconductor materials.

6.10 Case Studies

6.10.1 *Band Edge and Fermi Level Location*

A useful and important application of fs-UPS studies is the extraction of the Fermi level location at and near the semiconductor surface. An example is shown in Fig. 14 where spectra from both N- and P-type Si are compared. As mentioned before, the Fermi level of the system is determined by generating a photoelectron spectrum from a clean metal surface. That position is denoted as 0 eV in the spectrum. We expect a P-type semiconductor to exhibit a maximum in the valence band closer to the Fermi level while for an N-type semiconductor the valence band maximum will be further from the Fermi level as it will reside nearer to the conduction band edge of the semiconductor. The exact location is dependent upon the density of activated dopants in the material. Accurate determination of the Fermi level location within the band gap of a semiconductor requires that the bands at and near the surface are flat. Recall that for a typical semiconductor the bands at and near the surface are bent in response to surface dangling bonds, defects, and trace contaminants. As a result, standard UPS cannot be reliably used to determine the Fermi level location within the band gap. As described in this chapter, here again pumping the semiconductor with an intense pulse of light achieves the flat band condition that is needed to determine the Fermi level location within the band gap.

6.10.2 *The Valence Edge*

Before we start our measurement, it is important to assure that valence edge we are measuring is representative of the bulk and hence it is important to create an oxide-free and contaminant-free surface. In addition, a completely "clean" semiconductor surface may exhibit atomic rearrangement (reconstruction) that produce surface states within the gap; as a result, the valence edge measured in UPS will be a combination of the surface states and bulk valence states. In order to minimize the contributions from the surface, termination of surface dangling bonds will shift surface states to higher binding energy within the valence band. As an example, this can be achieved with hydrogen-termination or H-termination of the surface Si bonds. A simple method is to etch the Si surface in dilute HF acid that both removes SiO_2 and also H-terminates the surface. In the vacuum analysis system, a mild anneal to 150°–200°C removes surface adsorbed water leaving a clean, H-terminated surface. For other semiconductors, achieving a clean surface is achieved on a case-by-case basis.

From the spectra in Fig. 14, we clearly observe a difference between n- and p-type Si. Extrapolating the valence edge to zero intensity gives the location of the valence band maximum relative to the Fermi level [6, 7]. Consider the two cases in Fig. 14. Following removal of the native oxide from the surface of N-type Si with a phosphorous doping level of $5 \times 10^{15}/\text{cm}^3$, we observe that the valence band is 0.9 eV below the Fermi level. Since the band gap of Si is 1.1 eV at 300 K, this locates the Fermi level 0.2 eV below the conduction band minimum. The calculated location for the Fermi level in Si for this doping density is 0.23 eV, so measurement is within 30 meV of the calculated bulk value. In comparison for the P-type Si prepared similarly, with a boron doping level of $2 \times 10^{16}/\text{cm}^3$ we find the valence band maximum

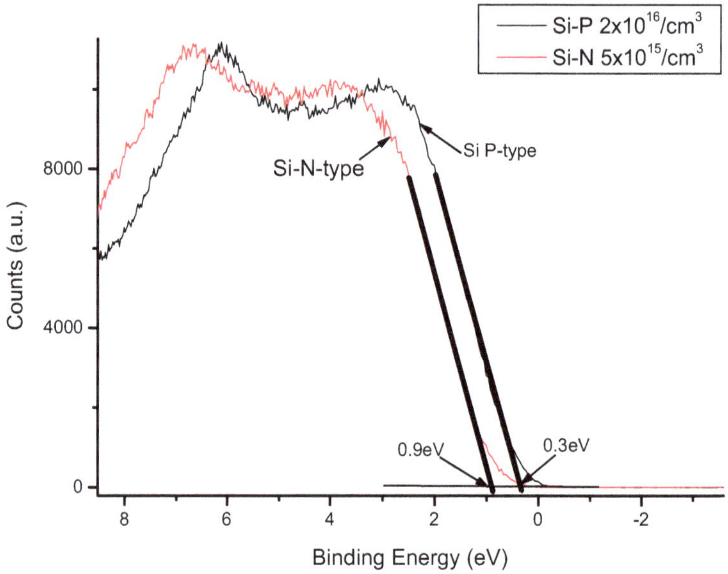

Fig. 14. Fs-UPS spectra from n-type and p-type Si samples. Under flat band conditions the valence band maximum locations are consistent with both the bulk doping densities and dopant type.

0.3 eV below the Fermi level a value that is ~80 meV different from the calculated value. Given that the actual doping levels quoted for commercial wafers possess some uncertainty these experimentally determined values from UPS are quite close to the calculated values for both doping types.

This example points to a useful application of this approach. For semiconductors, such as the multinaries used for thin-film photovoltaics that have essentially unknown doping levels, the Fermi level can be extracted with reasonable accuracy. In addition, one can work backwards, utilizing the Fermi level location to determine the density of activated dopants through the use of Eqs. (4) and (5). This is particularly important in complex and newly developed semiconductor materials for photovoltaics and other applications in energy, storage, memory, and more.

6.10.3 *High-k Dielectrics and Metal Gates*

The relentless scaling of transistors to ever smaller dimensions is essentially dictated by Moore's law [17] that calls for the doubling of device densities roughly every 18 months. When complementary metal–oxide–semiconductor (CMOS) channel lengths were of the order of microns or even many tens to hundreds of nanometers, scaling to smaller dimensions required sequential reductions in photomask features used in lithographic printing of Si wafers. No significant changes in the materials used to fabricate transistors at these length scales were necessary. The base semiconductor was and continues to be Si, the gate dielectric in devices was SiO_2 and the gate "metal" was heavily doped polycrystalline Si. SiO_2 was nature's gift to the semiconductor electronics industry — thermally stable, low defect density, possessing a sufficiently high dielectric constant. In addition, through the treatment with forming gas — a few % H_2 in Ar or N_2 gas at elevated temperatures, the density of defects at the SiO_2/Si interface could be reduced to $\sim 10^{10}/\text{cm}^2-$ eV. This low density of interfacial defects produced extremely high switching frequencies, excellent threshold voltage, and high I_{on}/I_{off} ratios, all required for producing efficient and reliable transistors used in computers. But as scaling continued to channel lengths below 20 nm, the decreasing thickness of SiO_2 began to exhibit unacceptable current leakage leading to significant power consumption.

6.10.4 *The High-k Dielectric*

One of the tenets of scaling, as delineated by Denard [18–20] required deliberate reductions of the gate oxide thickness with decreases in transistor dimension and in particular channel widths. Below ~20 nm channel widths, and given the modest (but previously sufficient) dielectric constant of SiO_2 [21] the thickness of the oxide layer resulted in increasing leakage leading to significant "off-state" power dissipation and diminishing on-off ratios. In other words, current leakage across the gate dielectric led to a condition where there was not much difference as to whether the transistor was on or off and even in the off state the device continued to dissipate power. This became intolerable for performance of the devices and had to be solved with the use of gate insulators with substantially higher dielectric constant — hence the term "high-k dielectrics".

Several candidate dielectrics included hafnium oxides and silicates, aluminum oxide, zirconium oxide, and more [21]. Employing higher-k dielectrics meant that physically thicker gate insulators could be used to dramatically reduce current leakage while maintaining the required capacitance. The lower current leakage would substantially improve power dissipation in both the on and off state. Replacement of SiO_2 is challenging since few dielectrics possess the physical and electrical properties of SiO_2. These properties include low interface state and bulk defect densities and stability at the high temperatures associated with device processing. These temperatures can reach as high as 1000°C, where interface reactions and atomic diffusion can occur, leading to the degradation of deposited layers. Particularly important is the potential formation of electrically active defects or traps which can impact electron mobility in the Si under the gate oxide and introduce threshold variations [9, 22]. These traps can be formed during device fabrication and under operating electrical stress.

Hafnium oxide (and silicates) [21] were a leading contender for replacing SiO_2 as the gate oxide in CMOS devices owing to its high dielectric constant ($\kappa \approx 21$) and relative ease of deposition. Unlike SiO_2, though, issues concerning bulk and interface defect densities that can lead to a pinning of the Fermi level within the Si band gap and high temperature instability, have hampered incorporation into CMOS devices.

Typical C–V analysis of defects in an MOS stack requires the fabrication of capacitors. This can often take weeks in a fab line to produce. Even more problematic is the study of the oxide/semiconductor system in the absence of a gate metal. This type of study is critical to understanding the evolution of a dielectric upon annealing or exposure to processing gases. To overcome this FPS can be utilized to study the properties of the oxide in its pristine state as well as following various processing steps [23]. As an example, and as shown in Fig. 15, as an oxide is annealed the formation of charged defects can occur. The impact of trapped charge within the oxide is to change the charge balance between the oxide and the semiconductor and this in turn leads to changes in the fields within the semiconductor and the resultant band bending. Hence, the band bending can be used as a sensor to measure changes in the density of charged defects within the oxide.

FPS was used to study the formation of defects at the Si/high-κ interface and inside the dielectric itself [4]. In these studies, the relationship between annealing temperature and the onset of oxide defect formation and charging was studied along with morphological changes in the oxide structure at elevated temperatures. A series of experiments were carried out on samples consisting of lightly doped $n-10^{14}/\text{cm}^3$ and $p-10^{15}/\text{cm}^3$ Si (100) substrates covered with 1 nm SiON and 1.2 nm HfO_2. Band bending values were measured after the samples were annealed at each temperature shown in Fig. 15 and then allowed to cool to room temperature. Temperatures ranged from 25°C to 950°C with solid circles representing the magnitude of the band bending. For *p*-Si (100) [Fig. 15(a)] we note a continuous decrease in the band bending up to 700°C, followed by a substantial change through 750°C into accumulation, wherein the bands, normally bent downward for *p*-type, are bent upward. Such behavior indicates that *negative* charge is driven into states in the dielectric stack. Also shown in Fig. 15(a) are data from SiON/Si

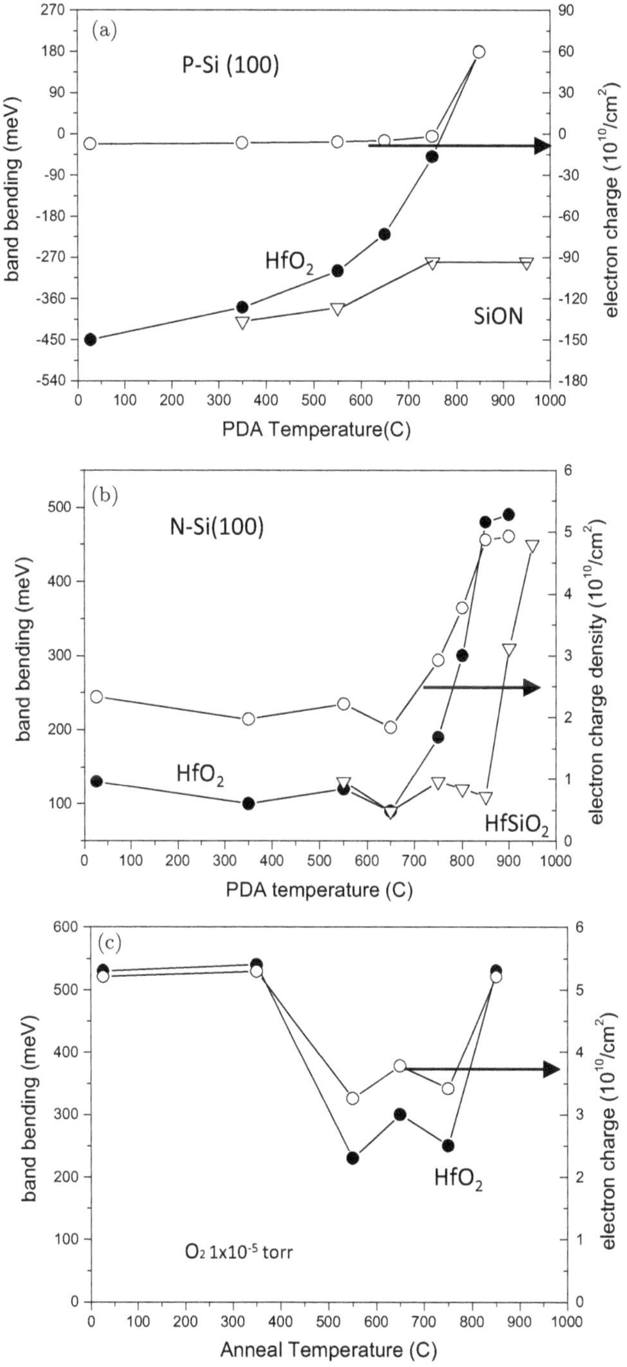

Fig. 15. (a) Band bending as a function of post-deposition anneal for HfO_2 and SiON on p-Si (100). (b) Band bending as a function of anneal temperature for HfO_2 and $Hf_{0.8}Si_{0.2}O_2$ on n-Si (100). (c) Band bending of HfO_2 on Si (100) during anneal exposed to 1×10^{-5} torr O_2 [23].

(100) (open triangles) without HfO_2. Unlike the HfO_2 covered system, the bands shift only 150 meV out to 950°C. This result indicates that the changes we observe occur predominantly in the HfO_2 layer and indicate the higher stability of the SiON layer relative to HfO_2. In Fig. 15(b), an analogous series is shown for *n*-type Si (100) with a dielectric stack consisting of 1 nm SiON and 1.5 nm HfO_2. Little change is observed in the band bending (solid circles) until an annealing temperature of 700°C is reached, beyond which an abrupt change of 400 meV in the band bending is observed. Note also that the direction of the band bending for both *p*- and *n*-type Si substrates is in response to the introduction of *negative* charge in the dielectric stack. Plotted as well in Fig. 15(b) (open triangles) is a series of measurements for $Hf_{0.8}Si0_{.2}O_2$ (hafnium silicate) on SiON grown *ex situ* by CVD on *n*-Si(100), which shows behavior similar to HfO_2 films, except that the abrupt change in band bending occurs 100°C higher. The incorporation of Si into the HfO2 structure modifies a number of properties including dielectric constant and a higher temperature at which the amorphous oxide converts to a polycrystalline structure. Using higher energy XUV harmonics (in this case 51.15 eV, the 33rd harmonic of 800 nm), Fig. 16 shows the evolution of HfO_2 deposited on a thin layer of SiON on Si. The O2s core level is identified in SiON and following the deposition of a 2.8 nm HfO_2 layer. Readily seen are the Hf 4f and 5p core levels. At temperatures between 800°C and 840°C we notice a splitting in the valence band at a binding energy of ~7 eV comprised of O2p and Hf 5d levels.

By 900°C we note significant state density up to the Fermi level that can be compared with pure Hf metal. Note the lower binding energy location of the Hf^{o}4f state in the metal relative to the Hf^{4+} location in the oxide.

The changes observed in the spectra of Fig. 16 are reflected in the transmission electron micrographs shown in Fig. 17. At 650°C the oxide is still amorphous (Fig. 17(a)) while at 870°C (Fig. 17(b)) we observe diffraction consistent with the formation of crystalline material and a thickening of the SiON layer associated with diffusion of O out from the HfO_2. This closely correlates with the dramatic increase in band bending in Si (see Figs. 15 (a) and 15(b)) and is consistent with a significant increase in the formation of O vacancies in the HfO_2.

The charging we observe is common to dielectric stacks grown on both *n*- and *p*-type substrates and manifests itself in all cases as a relatively abrupt and substantial change in the band bending at temperatures above 750°C. The amount of charge in the underlying Si, which can be calculated from the magnitude of the band bending is balanced by charge trapped at the Si/SiON interface (interface states), fixed charge within the SiON layer and

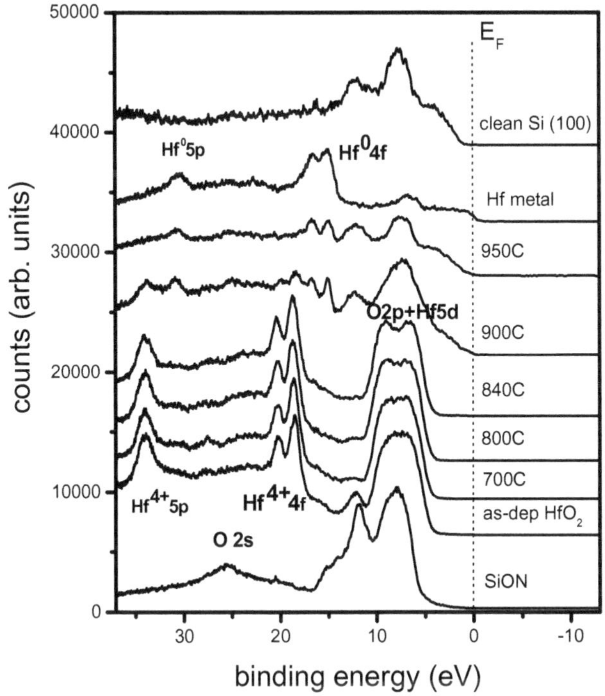

Fig. 16. Series of spectra from SiON and 2.8 nm HfO_2 deposited on Si (100) as a function of post-deposition anneal using the 33rd harmonic of 800 nm femtosecond light from an amplified Ti:sapphire laser [23].

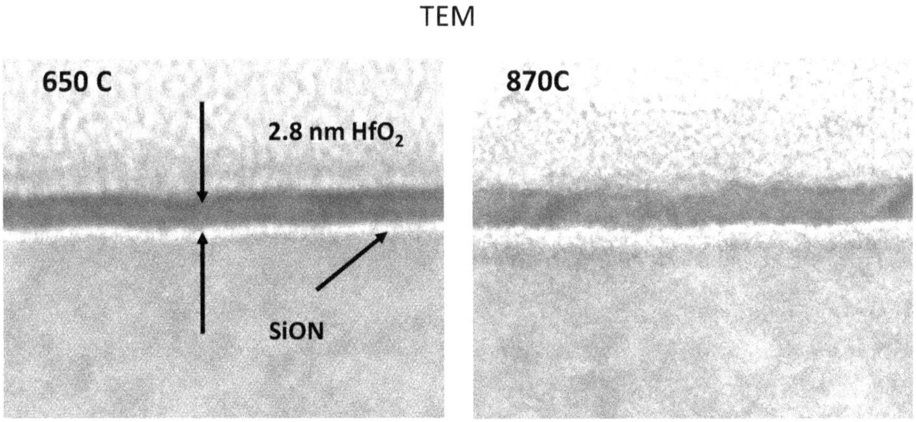

Fig. 17. Transmission electron micrographs of HfO_2/SiON/Si(100) after 650°C and 870°C anneals. At 870°C there is a notable thickening of the SiON layer and crystallization of the HfO2 layer [23].

charge injected into the HfO_2. This is shown in Eq. (8).

$$Q_{\mathrm{Si}} = -(Q_{\mathrm{intfc}} + Q_{\mathrm{SiON}} + Q_{\mathrm{HfO2}}) \tag{8}$$

Initially, at room temperature we observe an asymmetry in the degree of band bending between n- and p-type, i.e. the n-Si bands display 150 meV of bending while p-Si display bands are bent by 450 meV. Fixed positive charge in the SiON would produce such an asymmetry. At higher annealing temperatures, negative charge is injected into the HfO_2 and the Si bands bend in response. In addition, we expect changes to occur in the charge occupation of states at the Si/SiON interface as the Si bands bend and the Fermi level sweeps through Si band gap.

While all of these considerations conspire to make a quantitative determination of the amount of charge injected into the HfO_2 problematic, the results indicate that $\sim 10^{11} - 10^{12}$ electrons /cm^2 are injected into the HfO_2 at anneal temperatures above 750°C.

To investigate the stability of the trapped charge in the HfO_2 films the samples were annealed up to 750°C in UHV, cooled, and remeasured. No change in the band bending was observed. The samples were also left for 16 h in the UHV system and then remeasured; again no change in the band bending was observed. Both experiments indicate that once defects within the metal oxide were created and charged, they were highly stable. In an additional experiment, the reversibility of the charging was investigated by exposing an n-type sample to 1×10^{-5} torr O_2 for 5 min. The results are shown in Fig. 15(c). Before exposure, a band bending of 550 meV was measured which reduced to 200 meV until temperatures in excess of 750°C were attained. By 850°C the 550 meV band bending was recovered indicating that the oxygen passivated defects within the film but did not remove them. At temperatures sufficient to drive off the oxygen, the original defects were again negatively charged.

These experiments reveal the exceptional utility of FPS in extracting the details of the charge kinetics and correlating with the electronic and structural properties of the metal oxide in this case. But even more complex systems can be studied and we extend this to the case where a gate metal is deposited atop the dielectric to form a full MOS structure which is discussed in the next section.

6.10.5 *The Metal Gate/High-k Stack*

As described above, FPS can tell a great deal about the processing dependent formation of charge in the oxide alone. But of key interest to technologists responsible for creating functional devices is the behavior of complete metal/high-κ/Si stacks that operate as the gate for the device. Information on the location of the Fermi level at the Si/high-κ interface is critical since this "effective work function", or as is termed here, IFL tells us in effect how high a voltage is required to turn the device on and off. Since the power dissipation is given by $P = V^{*}_{\text{threshold}} I_{\text{on}}$ the lower the voltage required to turn the device on, the lower the power dissipation.

A number of metal/HfO_2/Si stacks were studied with metals that included Pt and Re (high work function), and TiN, Zr, and Hf (low work function). In general, a *p*-FET utilizes an *n*-type Si channel such that a high work function metal gate is required; for a high work function metal electrons will transit from the lower work function *n*-type Si into the higher work function metal thereby bending the Si bands up toward the Fermi level. For an *n*-FET the complementary nature requires a low work function metal gate, typically TiN. However, thermal processing of the high-κ stack drives the IFL to a mid-gap location [11].

A further complication involves metal-induced gap states (MIGS) [24–26] within the band gap of the dielectric as well as extrinsic states such as vacancies or dangling bonds that modify the effective work function Φ_m of the metals [12–14]. Therefore, it is important to understand how the work function of a metal, and its impact on band bending of the silicon substrate, is correlated with intrinsic MIGS states and interface or near interface defect states generated during thermal treatment. In these studies, FPS measurements were used to study the effects of thermal processing on metals deposited on HfO_2/SiON/Si substrates.

As mentioned above, studies were carried out on the band alignment of Si in HfO_2 high-κ, MOS structures with Pt ($\Phi_m = 5.6\,\text{eV}$), Re (5.0 eV), Hf (3.93 eV), and Zr (4.05 eV) metal overlayers, deposited at room temperature (RT). Metallic layers of high work function Pt, Re, and low work function Zr and Hf were grown *in situ* on CVD (chemical vapor deposition) grown HfO_2 (3 nm)/SiON (1 nm)/Si substrate with $n \sim 5 \times 10^{17}/\text{cm}^3$ ($p \sim 1 \times 10^{15}/\text{cm}^3$) doped Si (100).

Figure 18 shows the IFL positions obtained from FPS measurements for Pt, Re, Hf, and Zr on both *p*- and *n*-type substrates as a function of annealing temperature. The IFL for Pt is located ~200 meV above the VBM, and those for Hf and Zr are located ~300 meV below the CBM. These Fermi energies are located farther away from the band edges of silicon for ideal p-FET and n-FET operations.

Thermal annealing drives the IFLs further into midgap — a result of charging of the HfO_2 dielectric — this shift was confirmed with C–V

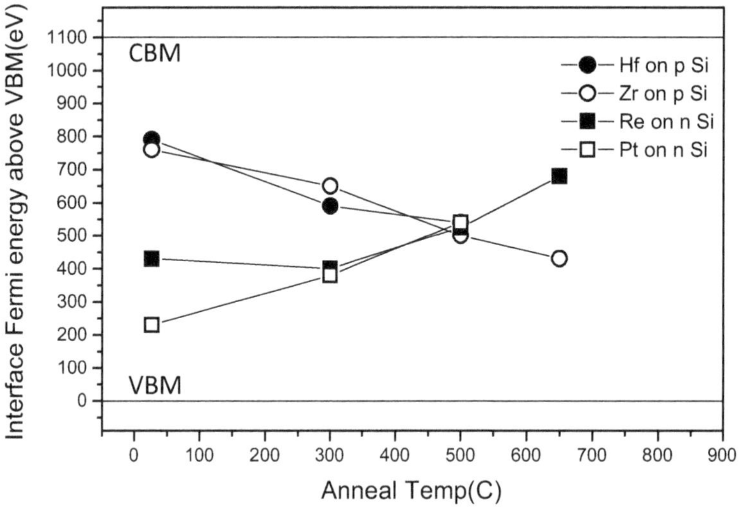

Fig. 18. IFL derived from FPS measurements for both *n*- and *p*-Si MOS stacks for the metals shown. Post-deposition anneals consistently drove the Fermi level to mid-Si gap [4].

measurements carried out *ex situ*. The same data are plotted as the effective work function in Fig. 19.

To calculate the effective work function of metals from the measured band bending $\mathrm{eV_{bb}}$, the voltage drop across the oxide, V_{ox} must be accounted for as shown in Eqs. (9) and (10).

$$\Phi_{\mathrm{eff}} = X^{\mathrm{Si}} + (E_{\mathrm{cbm}}^{\mathrm{bulk}} - E_{\mathrm{F}}^{\mathrm{bulk}}) + e\mathrm{V_{bb}} + e\mathrm{V_{ox}}, \tag{9}$$

$$V_{\mathrm{ox}} = V_{\mathrm{bb}}^{*} C_{\mathrm{Si}}/C_{\mathrm{ox}} + V_{\mathrm{ox}}^{\mathrm{interface}}. \tag{10}$$

Here, X^{Si} is Si electron affinity (4.1 eV), $E_{\mathrm{cbm}}^{\mathrm{bulk}}$ is the bulk conduction band minimum, and E_F^{bulk} is the Si bulk Fermi energy. C–V measurements of the density of states at the Si/SiON interface yield a value of at most $2 \times 10^{12}/\mathrm{cm}^2 - \mathrm{eV}$ distributed across the Si gap yielding a maximum 30 meV voltage drop across the SiON and HfO_2. The contribution from fixed charge in the SiON is small and ignored in the calculation [4]. The voltage drop across HfO_2 oxide due to the silicon space charge depends on the doping level and band bending of the substrate, and can be easily calculated from the ratio of the capacitances of silicon and oxide. For as-deposited samples, the effective work function values are $\Phi_{\mathrm{eff}} = 5.05\,\mathrm{eV} \pm 0.05$ (Pt), $4.85 \pm 0.05\,\mathrm{eV}$(Re), $4.40 \pm 0.05\,\mathrm{eV}$(Hf), and $4.43 \pm 0.05\,\mathrm{eV}$ (Zr). These values are quite different from vacuum work functions of the metals, but very close to the effective work functions calculated from the reported MIGS parameters

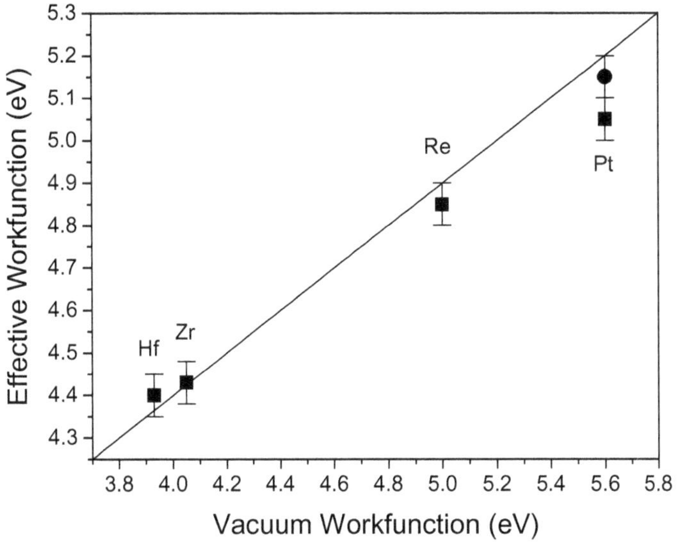

Fig. 19. Effective work function versus. vacuum work function. The effective work function determined as described in the text [4]. Solid squares are for as-deposited metals, solid circle (Pt) is following oxygen exposure.

[27, 28].

$$\Phi_{\text{eff}} = \Phi_{\text{cnl}} + \text{S}(\Phi_m - \Phi_{\text{cnl}}). \tag{11}$$

Here, Φ_{cnl} and S are the charge neutrality level and the Schottky pinning parameter of HfO_2. Using the reported values of $\Phi_{\text{cnl}} \sim 4.8\,\text{eV}$ from the vacuum level and $S \sim 0.5$, Φ_{eff} is 5.2 eV (Pt), 4.9 eV (Re), 4.35 eV (Hf), and 4.42 eV (Zr). Even though the contribution of extrinsic states or other defect states cannot be completely excluded, the calculation results agree well with experimental values of as-deposited samples (solid squares) as can be seen in Fig. 19, and intrinsic MIGS seems to be primarily responsible for reduced effective work function of the *as-deposited* samples. Plotted also in Fig. 19 is an oxygen-exposed Pt sample (solid circle) which is closer to the expected MIGs value. Oxygen plays an important role in determining the IFL position in high-κ/metal gate stacks.

Since it was determined that changes in the metal during anneal were minimal, the conclusion was that instabilities in the charged defect concentration within the oxide were critical. To investigate the impact of oxygen-related defects on midgap pinning of the IFL, Pt was employed as a gate metal. Pt is an ideal metal gate since it does not readily oxidize, it is stable at high temperatures, and O can easily diffuse through polycrystalline Pt, presumably along grain boundaries. Figure 20 shows the UPS spectra

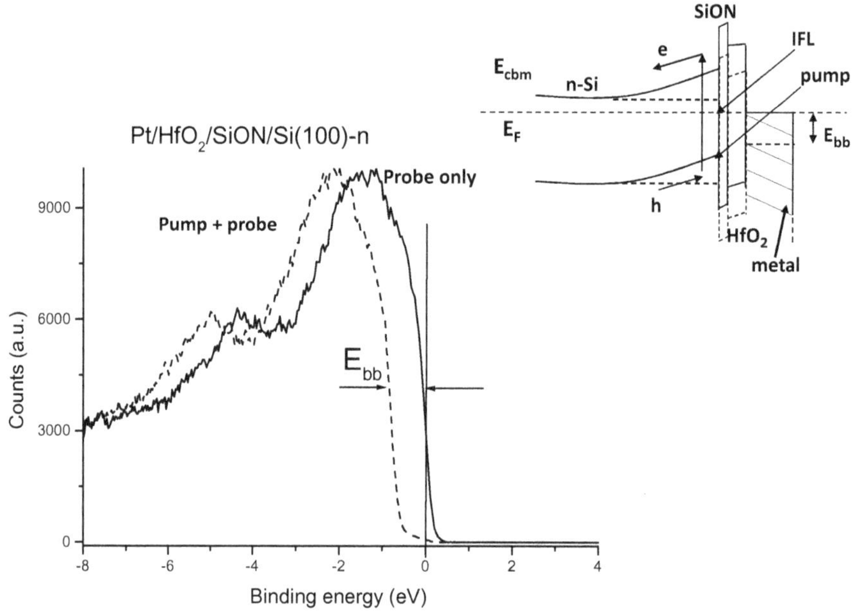

Fig. 20. Pt/HfO_2/SiON/Si(100)-n stack. Under photoexcitation the underlying Si bands flatten resulting in a shift of the photoelectron spectrum given as E_{bb}. This 0.5 eV shift corresponds to the band bending in the underlying Si and the direction of the shift from low to higher binding energy confirms the n-doped nature of the Si [4].

both before and after photoexcitation for a Pt/HfO_2/SiON/Si(100)-n stack. Also shown is the electronic structure derived from the FPS analysis and an extraction of 500 meV band bending, E_{bb}, of the underlying Si.

In Fig. 21, the as-deposited Pt produces an IFL 200 meV above the Si VBM which moves to the Si midgap as it is annealed. This change is determined simply by monitoring the change in the Si band bending following anneal; movement toward midgap indicates that the magnitude of the Si band bending is becoming smaller. After annealing to 500°C which produced a midgap IFL, the sample was then exposed to an O_2 ambient of 10 torr for 12 min at a temperature of 350°C in an attempt to unpin the Fermi level. Photoelectron spectra showed no sign of oxidation of Pt. As a result of the oxygen exposure the IFL was driven to within 130 meV of the Si VBM, a position even lower in the gap than for the as-deposited Pt. Subsequent annealing to 300° and 500°C reproduces the data points achieved in the annealing experiment before oxygen exposure, with the IFL again moving to midgap. A second round of oxygen exposure and annealing shows the remarkable cyclic reproducibility. The cyclic nature of the results suggests that the exposure to oxygen removes or passivates oxygen-derived defects; a

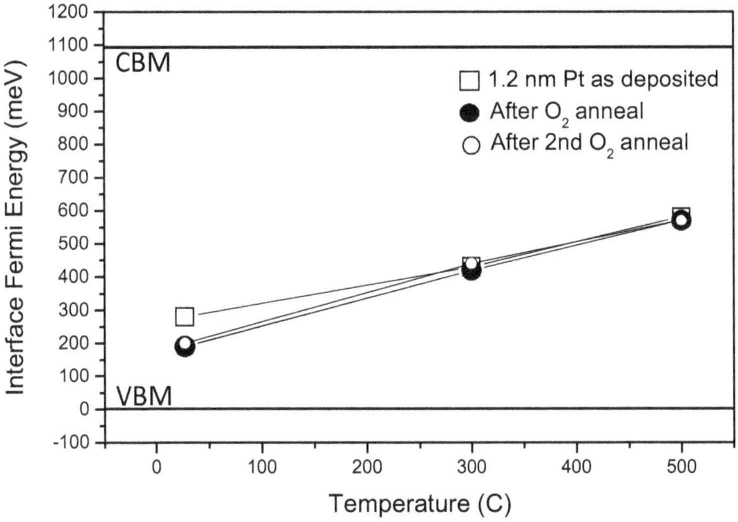

Fig. 21. Location of interface fermi level as an IFL for Pt (open boxes) on HfO2-SiON-Si (100)-n as a function of anneal temperature. Exposure to 10 Torr O2 (solid circles) followed by an anneal sequence, and a second cycle of O2 exposure (open circles) and anneal sequence is also shown [14].

model in which oxygen vacancies at the HfO_2/Pt interface are reduced during anneal and then oxidized during exposure is consistent with the femtosecond-photovoltage measurements.

These experiments indicate that, at least for high work function metals, the location of the IFL can be explained consistently with the formation of charged oxygen derived defect states in the HfO_2 or at its interface. Oxygen vacancies in hafnium oxide form donor-type defect states that are located energetically near to or within the silicon band gap [29, 30], can become positively charged when trapped electrons tunnel into lower lying, empty energy states within the high work function metal. The charge transfer will result in the formation of an electric dipole and reduced effective work function of the metal. In addition, oxygen vacancies at the interface between the gate metal and oxide can create interface states which produce a short-range interface dipole field, which in turn alters the effective gate metal work function and impact the band bending and hence IFL at the Si/oxide interface.

The experiments above show that FPS can be used to directly monitor the charge state of the metal/high-κ dielectric/semiconductor stack through changes in the underlying band bending of the semiconductor. This is a key advance in our ability to measure the properties of full-stacked films without the fabrication of capacitors, with a high degree of precision. Other systems under study include III–V materials such as InGaAs [31] that possess not only higher mobilities than Si, but also introduce higher complexity in the fabrication of such materials and their intrinsic interface oxides and defect states.

6.10.6 *Heterostructure Band Offsets: Application to Photovoltaics*

Heterostructures are a class of materials in which two dissimilar materials are coupled together. In the previous section, we described a specific multi-layer stack of materials (metal, oxide, and semiconductor) which is a specific class of heterostructure. The alignment of the valence and conduction bands and the location of the Fermi level within the materials and at the interface between dissimilar materials are critically important for the proper functioning of an electronic device.

The typical approach to identifying the alignment of bands of two dissimilar materials and the resultant band bending is to start with an atomically clean substrate and to measure the energetic location of atomic core levels of the substrate. Once established, the overlayer material is then sequentially deposited in layers sufficiently thin so as to permit observation of the core levels of the substrate; band bending is deduced by the energetic shifts with increasing overlayer thickness. Drawbacks of this approach include chemical shifts of the substrate core levels at the interface and the need to sequentially deposit the overlayer in the vacuum system without an air break. FPS spectroscopy avoids these issues and this approach is particularly useful when studying p-n junctions in photovoltaic heterostructures, where deposition of the thin n-layer on top of the p-absorber layer is carried out in a chemical bath in air and hence is not amenable to sequential vacuum system depositions.

Photovoltaics are another class of materials of great interest to industry and involve complex thin-film absorbers and their contacts. Given the state of global climate change, the need to move from fossil fuels as a primary source of energy to renewable technologies is of paramount importance. At the time of this writing, several technologies including solar, wind, and bio-fuels are making significant inroads into the energy generation landscape. Focusing on solar technologies, there are several approaches that include large area flat panel modules using single-junction approaches and solar concentration. Solar concentration itself involves different technologies — focusing of sunlight onto multi-junction III–V materials and solar thermal. Concentrated solar thermal involves the focusing of sunlight onto tubes containing high heat capacity molten salts; the stored thermal energy can then be used to run a steam turbine to generate electricity. This technology is additionally attractive in that stored thermal energy can be used to generate electricity even at night.

Focusing sunlight onto multi-junction materials at a concentration level of hundreds to thousands of suns (1 sun $\approx 1\,\mathrm{kW/m^2}$) results in conversion efficiencies of >45% of the sunlight into electricity [32]. Considering that the best lab device efficiencies for single crystal Si at 1 sun are ~25%, multi-junction solar cell performance is a spectacular increase. But in addition to the much higher production costs of multi-junction III–V solar cells, such a technology requires clear sunlit days since the sun is imaged onto the device and accurate tracking of the sun across the sky is required. Unfortunately, even a modestly hazy day results in dramatically reduced efficiency. Finally, energy is lost during DC to AC conversion (an issue with all PV technologies) and tracking of the sun.

In competition with the above technologies are single-junction solar cells. Most familiar is the Si solar panel found on home rooftops and commercial sites. Si is a widely available, earth abundant, and non-toxic semiconductor material whose polycrystalline single-junction performance is well above 20% leading to its wide use in solar-electricity generation [33]. Si suffers from several issues. It is an indirect band gap material that results in relatively weak light absorption — this requires Si absorber thicknesses to be ~200–300 microns. This leads to large material use, and relatively heavy panels. Also, refining Si into large area panels is itself energy intensive. On the positive side, Si has been utilized for many years in the electronics industry and as a result is one of the most well-understood materials known. As a result of its extensive study, Si-based PV has reached efficiency levels that makes it the go-to solar panel material.

Alternatives to Si are a class of absorbers utilized in "thin-film photovoltaics" or thin-film PV. Among the most-efficient thin-film absorber materials are GaAs, CdTe, CIGSe (Cu, In, Ga, Se), CZTS, Se (Cu, Zn, Sn, S, Se). The champion of thin-film absorbers is GaAs, a III–V direct band gap material that has achieved a single crystal efficiency of ~28% with absorber thicknesses of only 1 micron. But GaAs as well as other III-V materials will never be used in large area panels owing to their extremely high production costs, low earth abundance, arsenic toxicity, and competition for uses in the electronics industry for transistors, lasers, and lighting.

Other thin-film PV absorbers such as CdTe and CIGSe used in photovoltaic devices possess single-junction efficiencies above 20% and so are competitive with Si. As these materials possess direct band gaps (in addition to GaAs and other III–V materials discussed above) they absorb light within a depth of 1 micron. Hence, they require only 0.3% of the material compared with Si, and typically require much less energy to manufacture. They often can be deposited and processed on inexpensive substrates such as glasses and flexible high temperature polymers. The drawbacks include material scarcity[a]; Te, In, Ga, and Se are not found in abundance in the earth's crust and Te and Cd are toxic metals [34]. In and Ga find considerable use in the electronics industry, In for use in transparent conductors such as indium–tin–oxide (ITO) and both In and Ga are used in III–V electronics for transistors, solid state lasers, and lighting. There is enormous industrial competition for the available materials mined from the earth's crust.

PV absorbers such as the kesterites CZTS and CZTSSe [35–37] ($Cu_2ZnSn\ (S_xSe_{1-x})_4$) utilize materials that are abundant in the earth's crust, but to date PV devices fabricated from these materials have achieved only slightly more than half the efficiency of CdTe and CIGSe and less than half of GaAs. There are many other candidate materials, both earth abundant and not that are being pursued for higher efficiency solar cells. In particular, an explosion in the number of potential multinary compounds, through the use of combinatorial computational and experimental methods has led to a bewildering choice of materials.

A key to the high performance of these materials in PV devices is the choice of contact materials including both front and back contacts. Measurements of the electronic energy level alignments and the band bending created by these heterojunctions are of critical importance [38–41]. An example, as shown in Fig. 22, an effective heterojunction involves the deposition of an n-type contact material, also known as a buffer, in this case CdS, atop CZTSSe. Here, the CZTSSe is deposited on a Mo/glass substrate by spin coating a solution of precursors dissolved in hydrazine. The same CZTSSe can also be grown by vacuum deposition [42, 43]. In both cases, the films must be annealed at temperatures at and near 600°C in order to form a polycrystalline film consisting of large (1–2 μ) crystalline grains. Following formation of the p-type polycrystalline film, a thin layer of CdS is typically grown by a method called chemical bath deposition atop the CZTSSe film. The heterojunction formed by the n-type CdS and the p-type CZTSSe results in the transfer of electrons from the lower work function CdS into the higher work function CZTS.

The electron transfer bends the p-CZTS bands downward creating a depletion field that is critical to separating electrons and holes, created during the absorption of sunlight, and sweeping them efficiently to their

[a](http://minerals.usgs.gov/minerals/pubs/mcs/2015/mcs2015.pdf).

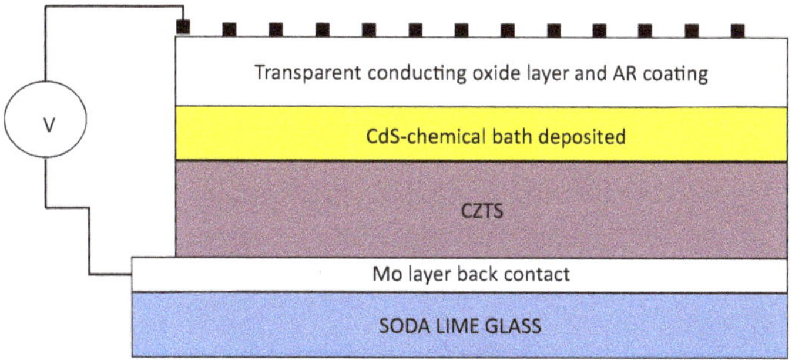

(A)

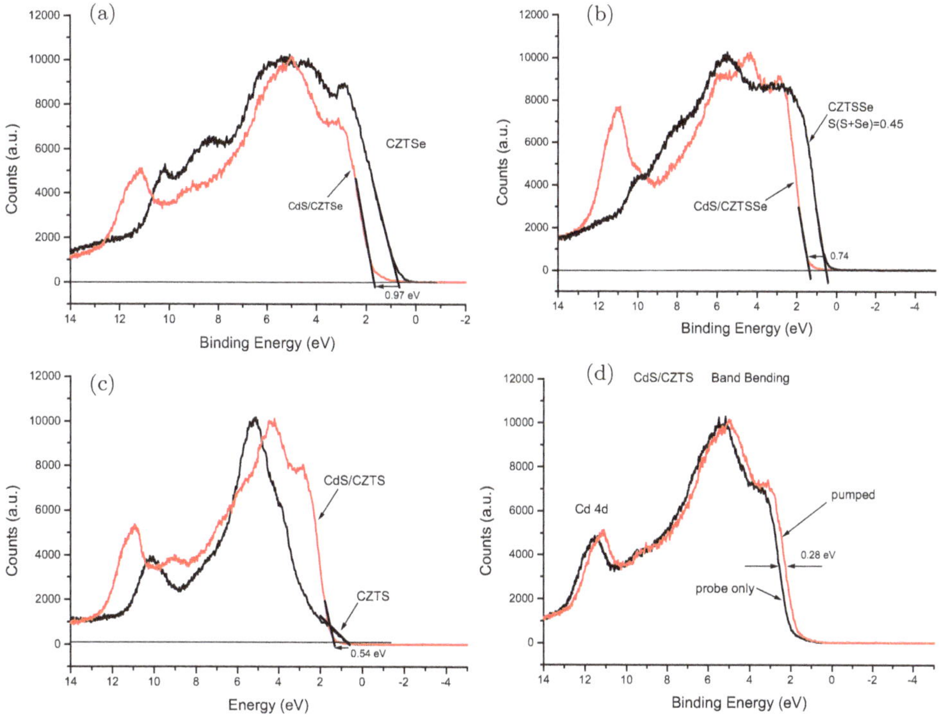

(B)

Fig. 22. (A) schematic of a thin-film photovoltaic cell. (B) Sequence of figures shows the comparison of pre- and post- chemical bath deposition of CdS atop (a)CZTSe, (b)CZTSSe and (c)CZTS. (d) Comparison of probe only and pumped CdS/CZTSe revealing a band bending of 0.28 eV for this heterostructure system. The direction of band bending identifies that CZTSe is *p*-type [38].

respective electrical contacts to create an electric current. The details of this were shown in Fig. 4. What makes FPS spectroscopy so important here is that, as described earlier, the CdS is deposited in a chemical bath and so the typical method of extracting band offsets through sequential depositions of overlayers without a vacuum break is not possible.

In the sequence of spectra in Fig. 22(a) corresponds to CdS/CZTSe (fully selenized), Fig. 22(b) to CdS/CZTSSe, Fig. 22(c) to CdS/CZTS (fully sulfurized) [38] and Fig. 22(d) displays the comparison of probe only and pumped spectra showing a 280 meV *p*-type band bending shift in the underlying CZTSe. This shift corresponds to the magnitude of the internal field so important to carrier separation that efficiently drives electrons into the CdS and holes toward the rear of the PV device.

Also note in Fig. 22 that the CdS and CZTSSe valence band edges must be determined accurately in the absence of band bending. If we recall from Section 6.10.2, this requires that the starting surface is free from oxides so that the CZTSSe surface valence band edge can be accurately determined. This was achieved through etching in dilute NH_4OH to remove residual oxide. Because of the complex polycrystalline and multi-elemental nature of this material, careful control of the overall processing of the material inside the vacuum analysis system was also crucial. Mild anneals below 200°C were required to remove residual adsorbed atmospheric water but higher anneal temperatures led to evaporation of S and Se into the vacuum that also modified the valence edges. Even gentle sputtering with Ar or Ne to achieve a clean surface is problematic as chalcogen sputtering rates are much higher than for the metals in the film.

Figure 23 provides a schematic of the valence, conduction band, and Fermi level locations allowing for the heterojunction electronic structure of the three different cases to be fully determined. While the valence band locations are measured via UPS, the conduction bands are determined through optical absorption determination of the band gaps of the CZTS, Se, and CdS separately.

In all cases, the CdS forms a beneficial “spike” in the conduction band offsets, i.e. the conduction band of the CdS is slightly higher than that of the CZTS. At first blush it seems somewhat counterintuitive that a small barrier to electron flow would be beneficial. If we consider the opposite case where the conduction band of the CdS is lower than that of the CZTS (called a “cliff”) we find that this results in a lower open circuit voltage for the device [44]. Since the power efficiency of a photovoltaic device depends on both the open circuit voltage, V_{oc}, that can be achieved as well as it short circuit current, J_{sc}, any property that lowers the voltage is a problem. The second

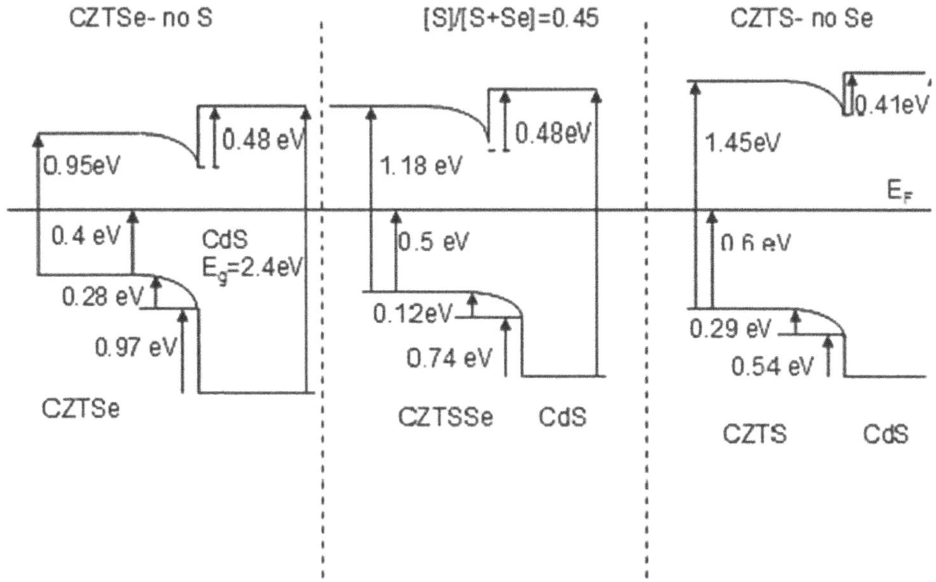

(a)

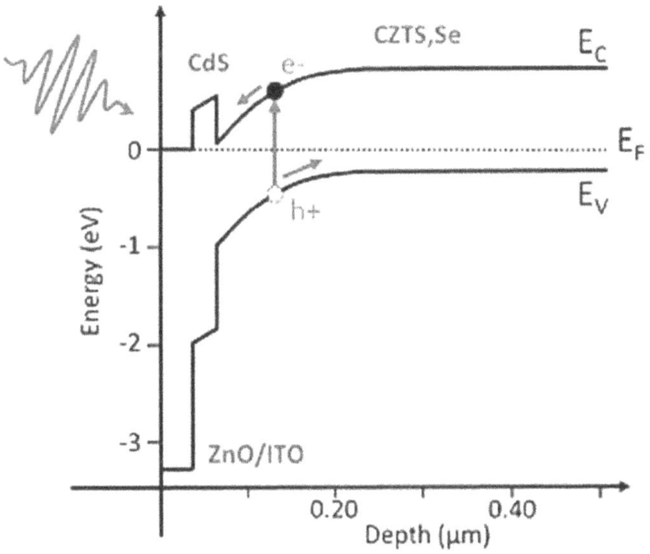

(b)

Fig. 23. (a) schematic energy band alignments as derived from FPS experiments for CdS/CZTSe, CZTS, Se, and CZTS. [38]. (b) Schematic used in WXamps calculations showing energy versus depth for the full ZnO/ITO/CdS/CZTS, Se stack.

reason for a spike offset is that it rapidly removes electrons from the region of the interface which typically possesses defects that enhance non-radiative recombination, robbing the device of both current and voltage. Of course, it is important that the magnitude of the spike is limited to below 0.5 eV otherwise current flow is pinched off as in the case when CdS is replaced by ZnS as a buffer where the ~1 eV spike reduces current in the device to zero (Fig. 24 middle) [40]. Since we want to collect only electrons at this junction it is important for the bands in the PV absorber to bend downward toward the *n*-type buffer, and for the valence offset to be large enough to present a significant barrier to hole transport into the buffer. The direction of the field also drives the electrons into the transparent conducting oxide where it is then collected by the grid of metal lines. For CdS/CZTS, Se these criteria are appropriately satisfied and as a result CdS is an optimal *n*-type buffer material for both CZTS Se and CIGSe.

The valence and conduction band lineups are critically important. As an example of this, we show in Fig. 24 several cases of buffer lineups with the PV absorber CZTSSe. These are pump/probe FPS measurements in which the photoelectron spectra of the clean CZTSSe surface is collected and then compared with spectra from samples with overlayer buffers of CdS, ZnO, ZnS, and In_2S_3. In all cases, the buffers are full or close to full thickness, typically grown by chemical bath deposition or atomic layer deposition. Using the FPS measurements, the valence electronic structures are measured and the locations of the conduction band minima were identified through optical absorption spectra. Typically the band gap of the CZTSSe was obtained through a method called EQE [45], or external quantum efficiency in which the full solar cell device is probed electronically while being exposed to light scanned typically from ~300 nm to 1500 nm. The electrical response of the device during this light exposure is used to determine the band gap [46].

As can be seen in Fig. 24 the valence electron spectra are collected for the unpumped and pumped heterostructure. This permits the extraction of the band bending by the rigid shift of the spectra before and after photoexcitation (see Fig. 22). The location of the valence band edges are then determined under flatband conditions.

The electronic structures including band bending can then be assembled as shown in Fig. 24. CdS/CZTSSe valence and conduction offsets (Fig. 23) are favorable for efficient device performance. This is similarly true for the In_2S_3/CZTSSe heterojunction. The ZnO/CZTSSe is an intermediate case where the conduction offset is nearly zero. There is no barrier but non-radiative recombination may be an issue for this heterojunction. For ZnS,

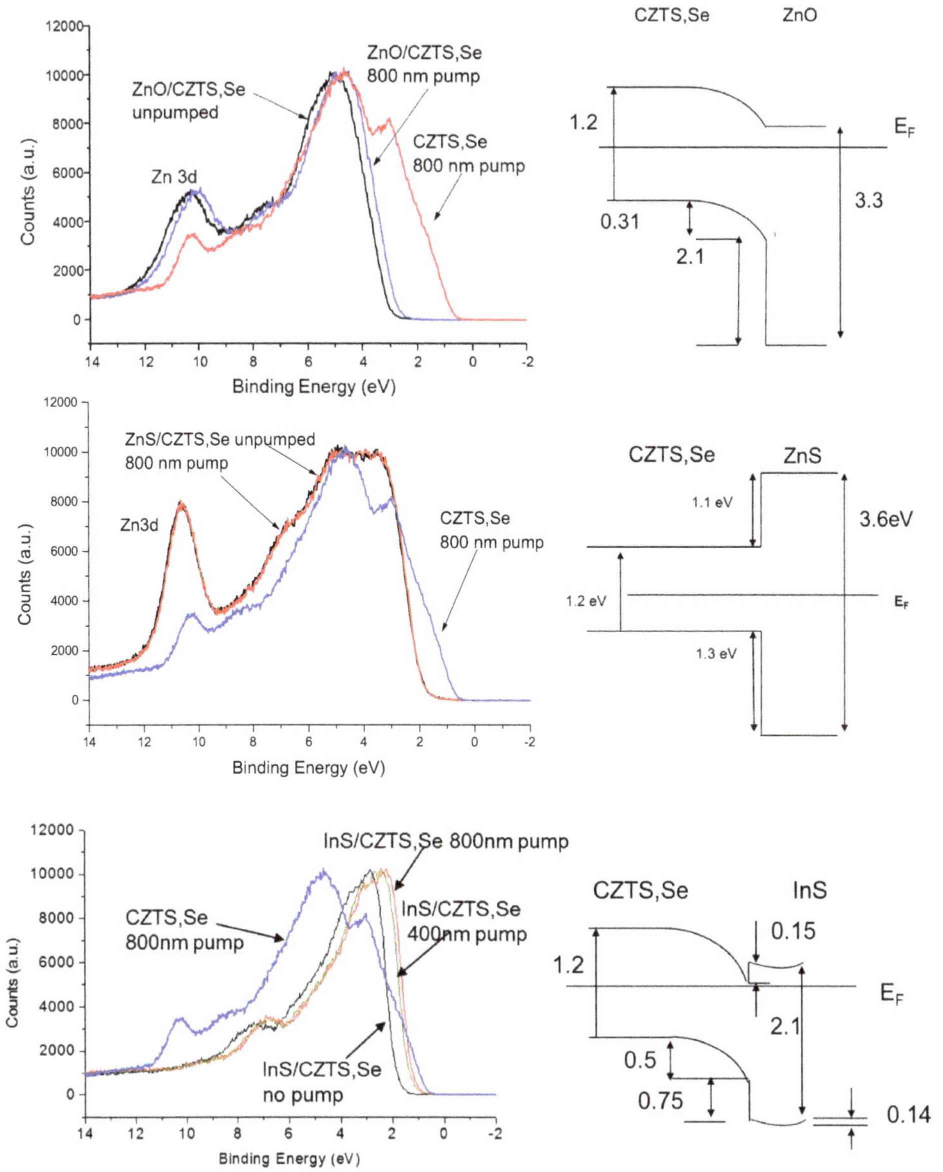

Fig. 24. FPS of top: ZnO/CZTS, Se, middle: ZnS/CZTSSe, bottom: In_2S_3/CZTSSe and their respective energy schematics. J–V curves for devices consisting of ZnO, ZnS, In_2S_3, and CdS buffers on CZTSSe. Note that the large conduction band offset for ZnS/CZTSSe results in zero current flow in the device as discussed in the text [40].

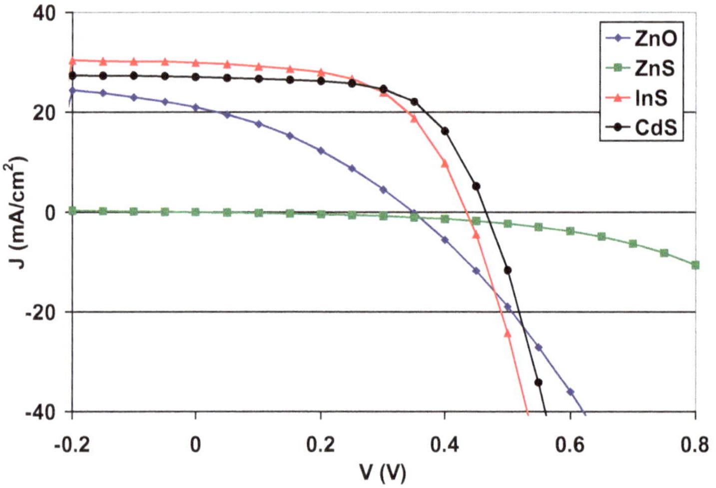

Fig. 24. (*Continued*)

the conduction barrier is >1 eV giving rise to a considerable barrier to current flow.

We may ask how these results compare with device performance. Shown in Fig. 24 just above are J–V curves generated in a standard PV device solar simulator and is the key measurement protocol for evaluating the short circuit current J_{sc}, open circuit voltage, V_{oc}, and other key solar cell performance parameters. As can be seen there is a strong correspondence between our barrier measurements and the device performance. For the CdS and In_2S_3 buffers, we observe high performance where the short circuit currents are high, >30 mA/cm^2. For ZnO the performance is intermediate, wherein the J_{sc} and V_{oc} are both lower. As was mentioned, the lack of a conduction band barrier at the heterojunction can lead to enhanced e-h recombination at the interface lowering both current and voltage, as seen in the device performance. Finally, the ZnS buffer is the worst performer. The large ZnS conduction band barrier of >1 eV means that electrons cannot transit through the barrier into the transparent conducting oxides and metal grid for current collection. Indeed, we observe no current flow in the device measurements for the ZnS/CZTS, Se device confirming that the FPS measurements provide the kind of key electronic structural information critical to device performance.

In this chapter, we introduced and described the concepts associated with FPS. Our focus was on the study of complex material stacks and devices that are of specific industrial use; metal–oxide–semiconductors, high-κ dielectrics, and photovoltaic heterojunctions and devices. We showed the methods for extracting the electron energy levels and internal dipole fields buried under layers of materials in realistic industrially relevant devices. These are but a few of the multitude of systems that are of keen interest both of presently used devices and those envisioned for the future.

References

[1] W. H. Brattain and J. Bardeen, "Surface properties of germanium," *Bell Syst. Tech. J.*, vol. 32, no. 1, pp. 1–41, 1953.

[2] H. C. Gatos and J. Lagowski, "Surface photovoltage spectroscopy — A new approach to the study of high-gap semiconductor surfaces," *J. Vac. Sci. Technol.*, vol. 10, no. 1, pp. 130–135, 1973.

[3] L. Kronik and Y. Shapira, "Surface photovoltage spectroscopy of semiconductor structures: At the crossroads of physics, chemistry and electrical engineering," *Surf. Interface Anal.*, vol. 31, no. 10, pp. 954–965, 2001.

[4] D. Lim and R. Haight, " *in situ* photovoltage measurements using femtosecond pump-probe photoelectron spectroscopy and its application to metal–HfO2–Si structures," *J. Vac. Sci. Technol. A*, vol. 23, no. 6, pp. 1698–1705, 2005.

[5] S. M. Sze and K. K. Ng, *Physics of Semiconductor Devices.* John wiley & sons, Hoboken, 2006.

[6] E. A. Kraut, R. W. Grant, J. R. Waldrop, and S. P. Kowalczyk, "Precise determination of the valence-band edge in X-ray photoemission spectra: Application to measurement of semiconductor interface potentials," *Phys. Rev. Lett.*, vol. 44, no. 24, p. 1620, 1980.

[7] S. A. Chambers, T. Droubay, T. C. Kaspar, and M. Gutowski, "Experimental determination of valence band maxima for SrTiO3, TiO2, and SrO and the associated valence band offsets with Si (001)," *J. Vac. Sci. Technol. B*, vol. 22, no. 4, pp. 2205–2215, 2004.

[8] D. Kahng, "A historical perspective on the development of MOS transistors and related devices," *IEEE Trans. Electron Devices*, vol. 23, no. 7, pp. 655–657, 1976.

[9] A. Kerber, E. Cartier, L. Pantisano, R. Degraeve, T. Kauerauf, Y. Kim, A. Hou, G. Groeseneken, H. E. Maes, and U. Schwalke, "Origin of the threshold voltage instability in SiO_2/HfO_2 dual layer gate dielectrics," *IEEE Electron Device Lett.*, vol. 24, no. 2, pp. 87–89, 2003.

[10] E. P. Gusev, E. Cartier, D. A. Buchanan, M. Gribelyuk, M. Copel, H. Okorn-Schmidt, and C. D'emic, "Ultrathin high-K metal oxides on silicon: Processing, characterization and integration issues," *Microelectron. Eng.*, vol. 59, no. 1, pp. 341–349, 2001.

[11] A. Callegari, E. Cartier, M. Gribelyuk, H. F. Okorn-Schmidt, and T. Zabel, "Physical and electrical characterization of Hafnium oxide and Hafnium silicate sputtered films," *J. Appl. Phys.*, vol. 90, no. 12, pp. 6466–6475, 2001.

[12] S. Zafar, A. Kumar, E. Gusev, and E. Cartier, "Threshold voltage instabilities in high-κ gate dielectric stacks," *IEEE Trans. Device Mater. Reliab.*, vol. 5, no. 1, pp. 45–64, 2005.

[13] E. Cartier, F. R. McFeely, V. Narayanan, P. Jamison, B. P. Linder, M. Copel, V. K. Paruchuri, V. S. Basker, R. Haight, and D. Lim, "Role of oxygen vacancies in V FB/V t stability of pFET metals on HfO 2," in *Digest of Technical Papers. 2005 Symposium on VLSI Technology,* 2005, pp. 230–231.

[14] D. Lim, R. Haight, M. Copel, and E. Cartier, "Oxygen defects and Fermi level location in metal–hafnium oxide–silicon structures," *Appl. Phys. Lett.*, vol. 87, no. 7, p. 72902, 2005.

[15] E. O. Johnson, "Large-signal surface photovoltage studies with germanium," *Phys. Rev.*, vol. 111, no. 1, p. 153, 1958.
[16] J. P. Long, H. R. Sadeghi, J. C. Rife, and M. N. Kabler, "Surface space-charge dynamics and surface recombination on silicon (111) surfaces measured with combined laser and synchrotron radiation," *Phys. Rev. Lett.*, vol. 64, no. 10, p. 1158, 1990.
[17] R. R. Schaller, "Moore's law: Past, present and future," *IEEE Spectr.*, vol. 34, no. 6, pp. 52–59, 1997.
[18] W. Haensch, E. J. Nowak, R. H. Dennard, P. M. Solomon, A. Bryant, O. H. Dokumaci, A. Kumar, X. Wang, J. B. Johnson, and M. V Fischetti, "Silicon CMOS devices beyond scaling," *IBM J. Res. Dev.*, vol. 50, no. 4.5, pp. 339–361, 2006.
[19] G. Baccarani, M. R. Wordeman, and R. H. Dennard, "Generalized scaling theory and its application to a 1/4 micrometer MOSFET design," *IEEE Trans. Electron Devices*, vol. 31, no. 4, pp. 452–462, 1984.
[20] D. J. Frank, R. H. Dennard, E. Nowak, P. M. Solomon, Y. Taur, and H.-S. P. Wong, "Device scaling limits of Si MOSFETs and their application dependencies," *Proc. IEEE*, vol. 89, no. 3, pp. 259–288, 2001.
[21] J. Robertson, "High dielectric constant gate oxides for metal oxide Si transistors," *Reports Prog. Phys.*, vol. 69, no. 2, p. 327, 2005.
[22] G. D. Wilk, R. M. Wallace, and J. M. Anthony, "High-κ gate dielectrics: Current status and materials properties considerations," *J. Appl. Phys.*, vol. 89, no. 10, pp. 5243–5275, 2001.
[23] D. Lim and R. Haight, "Temperature dependent defect formation and charging in hafnium oxides and silicates," *J. Vac. Sci. Technol. B*, vol. 23, no. 1, pp. 201–205, 2005.
[24] J. Tersoff, "Schottky barrier heights and the continuum of gap states," *Phys. Rev. Lett.*, vol. 52, no. 6, p. 465, 1984.
[25] J. Tersoff, "Theory of semiconductor heterojunctions: The role of quantum dipoles," *Phys. Rev. B*, vol. 30, no. 8, p. 4874, 1984.
[26] W. Mönch, "Barrier heights of real Schottky contacts explained by metal-induced gap states and lateral inhomogeneities," *J. Vac. Sci. Technol. B*, vol. 17, no. 4, pp. 1867–1876, 1999.
[27] J. Robertson, "Band offsets of wide-band-gap oxides and implications for future electronic devices," *J. Vac. Sci. Technol. B*, vol. 18, no. 3, pp. 1785–1791, 2000.
[28] Y.-C. Yeo, P. Ranade, T.-J. King, and C. Hu, "Effects of high-/spl kappa/gate dielectric materials on metal and silicon gate workfunctions," *IEEE Electron Device Lett.*, vol. 23, no. 6, pp. 342–344, 2002.
[29] K. Xiong, P. W. Peacock, and J. Robertson, "Defect Energy Levels in HfO2, ZrO2, La2O3, and SrTiO3," in *MRS Symposia Proceedings*, 2004, no. 811, p. D641.
[30] A. S. Foster, F. L. Gejo, A. L. Shluger, and R. M. Nieminen, "Vacancy and interstitial defects in hafnia," *Phys. Rev. B*, vol. 65, no. 17, p. 174117, 2002.
[31] A. Carr, J. Rozen, M. M. Frank, T. Ando, E. A. Cartier, P. Kerber, V. Narayanan, and R. Haight, "Evolution of interfacial Fermi level in In0. 53Ga0. 47As/high-κ/TiN gate stacks," *Appl. Phys. Lett.*, vol. 107, no. 1, p. 12103, 2015.
[32] M. A. Green, K. Emery, Y. Hishikawa, W. Warta, and E. D. Dunlop, "Solar cell efficiency tables (version 48)," *Prog. Photovoltaics Res. Appl.*, vol. 24, no. 7, pp. 905–913, 2016.
[33] P. J. Cousins, D. D. Smith, H.-C. Luan, J. Manning, T. D. Dennis, A. Waldhauer, K. E. Wilson, G. Harley, and W. P. Mulligan, "Generation 3: Improved performance at lower cost," in *Photovoltaic Specialists Conference (PVSC), 2010 35th IEEE*, 2010, pp. 275–278.

[34] C. Candelise, M. Winskel, and R. Gross, "Implications for CdTe and CIGS technologies production costs of indium and tellurium scarcity," *Prog. Photovoltaics Res. Appl.*, vol. 20, no. 6, pp. 816–831, 2012.

[35] H. Katagiri, K. Jimbo, W. S. Maw, K. Oishi, M. Yamazaki, H. Araki, and A. Takeuchi, "Development of CZTS-based thin film solar cells," *Thin Solid Films*, vol. 517, no. 7, pp. 2455–2460, 2009.

[36] H. Katagiri, "Cu_2 $ZnSnS_4$ thin film solar cells," *Thin Solid Films* vol. 480, pp. 426–432, 2005.

[37] W. Wang, M. T. Winkler, O. Gunawan, T. Gokmen, T. K. Todorov, Y. Zhu, and D. B. Mitzi, "Device characteristics of CZTSSe thin-film solar cells with 12.6% efficiency," *Adv. Energy Mater.*, vol. 4, no. 7, pp. 1–5, 2014.

[38] R. Haight, A. Barkhouse, O. Gunawan, B. Shin, M. Copel, M. Hopstaken, and D. B. Mitzi, "Band alignment at the Cu2ZnSn(SxSe1-x)4/CdS interface," *Appl. Phys. Lett.*, vol. 98, no. 25, p. No pp yet given, 2011.

[39] L. Sun, R. Haight, P. Sinsermsuksakul, S. Bok Kim, H. H. Park, and R. G. Gordon, "Band alignment of SnS/Zn(O,S) heterojunctions in SnS thin film solar cells," *Appl. Phys. Lett.*, vol. 103, no. 18, 2013.

[40] D. A. R. Barkhouse, R. Haight, N. Sakai, H. Hiroi, H. Sugimoto, and D. B. Mitzi, "Cd-free buffer layer materials on Cu 2ZnSn(S xSe 1-x) 4: Band alignments with ZnO, ZnS, and In 2S 3," *Appl. Phys. Lett.*, vol. 100, no. 19, pp. 4–9, 2012.

[41] M. Bär, S. Nishiwaki, L. Weinhardt, S. Pookpanratana, W. N. Shafarman, and C. Heske, "Electronic level alignment at the deeply buried absorber/Mo interface in chalcopyrite-based thin film solar cells," *Appl. Phys. Lett.*, vol. 93, no. 4, p. 2110, 2008.

[42] I. Repins, C. Beall, N. Vora, C. DeHart, D. Kuciauskas, P. Dippo, B. To, J. Mann, W.-C. Hsu, and A. Goodrich, "Co-evaporated Cu_2 $ZnSnSe_4$ films and devices," *Sol. Energy Mater. Sol. Cells*, vol. 101, pp. 154–159, 2012.

[43] Y. S. Lee, T. Gershon, O. Gunawan, T. K. Todorov, T. Gokmen, Y. Virgus, and S. Guha, "Cu<inf>2</inf> ZnSnSe<inf>4</inf>thin-film solar cells by thermal co-evaporation with 11.6% efficiency and improved minority carrier diffusion length," *Adv. Energy Mater.*, vol. 5, no. 7, pp. 2–5, 2015.

[44] T. Minemoto, Y. Hashimoto, T. Satoh, T. Negami, H. Takakura, and Y. Hamakawa, "Cu(In,Ga)Se[sub 2] solar cells with controlled conduction band offset of window/Cu(In,Ga)Se[sub 2] layers," *J. Appl. Phys.*, vol. 89, no. 12, p. 8327, 2001.

[45] T. Gokmen, O. Gunawan, T. K. Todorov, and D. B. Mitzi, "Band tailing and efficiency limitation in kesterite solar cells," *Appl. Phys. Lett.*, vol. 103, no. 10, pp. 1–12, 2013.

[46] K. F. Tai, T. Gershon, O. Gunawan, and C. H. A. Huan, "Examination of electronic structure differences between CIGSSe and CZTSSe by photoluminescence study," *J. Appl. Phys.*, vol. 117, no. 23, p. 235701, 2015.

Chapter 7

Survey of Ultrafast Light Sources, Applications, and Final Thoughts

7.1 Introduction

In addition to femtosecond and even attosecond (10^{-18} s) light sources that we have described in the previous chapters, there are other novel light sources such as synchrotrons, free-electron lasers (FEL), and alternative design lasers that are operational, or soon to be on line that extend the wavelength regime to deep within the X-ray region of the spectrum. These light sources, as we will briefly describe, may not be presently used for solving industrial problems but may, in the near future, be used in that environment. Extreme UV (EUV) lithography operating at 13 nm wavelength is an example [1, 2]. High-energy pulsed CO_2 lasers generate intense plasmas in Sn droplets; the emitted radiation is collected, wavelength filtered and directed into an optical system used in patterning for semiconductor device manufacturing. While the CO_2 laser is not particularly short pulsed, the metrology tools that can be used to evaluate and examine the lithography quality can utilize high harmonics from femtosecond lasers.

Other unique light sources include X-ray FELs operating in the US, Japan, and in Europe [3] that produce femtosecond pulses of light at wavelengths down to 1–5 Å at repetition rates ranging from 10 to several thousand pulses per second at high pulse intensities. Some of these FELs include the LCLS at Stanford, FLASH and XFEL in Germany, FERMI and SPARX in Italy, PAL XFEL in Korea, and XFEL/Spring-8 in Japan, to name a few. New synchrotron facilities such as the NSLSII at Brookhaven National Lab are extraordinarily high brightness sources with pulsewidths in the picosecond regime and extraordinarily broad tunability that can span from the infrared to several hundred keV.

Facilities such as these, coupled with table-top soft X-ray high harmonic sources, open new vistas for imaging, materials characterization, and even manufacturing that must be considered. We will briefly touch on these new technologies and their impact on potential industrial utilization.

7.2 High-Harmonic Soft X-rays for Metrology

As described in earlier chapters, amplified ultrafast femtosecond lasers can produce substantial fluxes of highly coherent high-harmonic light in the soft X-ray regime [4–6]. These short wavelengths are well suited for nanoscale imaging of features. This can be shown via Abbe's criteria, similar to Rayleigh's, for the fundamental resolution of a microscope

$$r = \frac{\lambda}{2n \sin \theta},$$

where r is the resolution, λ is the wavelength of the incident light, n is the refractive index, and θ is the half angle of the light collected by the imaging lens. Wavelengths in the EUV and X-ray regime can be employed in suitably designed microscopes to achieve nanometer scale resolution. This is attractive in the industrial setting for imaging of nanomachined features, defect and mask inspection for semiconductor patterning and lithography, and robust imaging of biological samples. In applying this microscopy practically in the industrial setting, robustness and compactness of the imaging apparatus are required.

X-ray microscopes installed at high photon flux facilities such as synchrotrons employ special optics such as Fresnel zone plates for focusing of energetic light to small spots. Fresnel zone plates utilize diffraction of incident X-rays through a thin Si element etched to create a series of radially symmetric concentric rings. Constructive interference of the diffracted X-ray light transmitted through the open regions of the zone plate produces a small focal spot on the sample being studied. A primary drawback of a zone plate is that transmission of the X-rays is lossy, so that high intensities are needed. The plates are somewhat difficult to manufacture due to the precision required to yield high resolution images with minimal aberrations and they tend to have very short focal lengths that limit imaging capabilities. Other focal options include the use of either grazing incidence toroidal or ellipsoidal mirrors and multilayer coated normal incidence focusing mirrors that typically work on the soft-X-ray regime.

An emerging alternative method of imaging with X-rays is called coherent diffractive imaging (CDI). Shown schematically in Figure 1, this approach does not utilize a focusing element but instead develops an image from the diffraction of incident light from the features on the sample. The image is reconstructed through computational algorithms [8]. In CDI, the

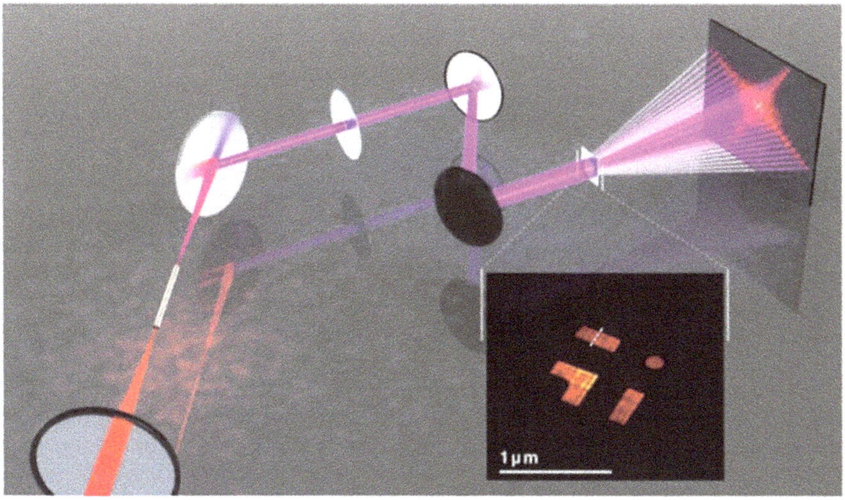

Fig. 1. Schematic layout of a femtosecond lensless CDI system. Coherent femtosecond EUV pulses are transmitted (in this example) through a patterned thin sample. The transmitted diffraction pattern is converted to a real-space picture through algorithms mentioned in the text. Adapted from reference [9].

diffraction pattern from monochromatic light incident on a sample is collected, and the Fourier amplitudes (light intensities) over the detector are recovered. A phase retrieval algorithm is then applied iteratively to calculate the corresponding phases, reconstructing a real-space image via Fourier inversion. An implementation well suited for industry involves utilizing the high harmonics of an amplified femtosecond laser as the incident source instead of a high-flux larger scale facility. This process has been demonstrated in both reflection and transmission geometries with resolution down to 22 nm using 13 nm light [9]. This approach has also been employed to image yeast spores, viruses, features on photomasks, nanoparticles, and more.

Since EUV and soft X-ray light can be routinely generated with femtosecond laser high harmonics, metrological applications for EUV lithography is applicable in industrial settings. Given the high spatial and temporal coherence of this light, analysis of optical components at 13 nm to look for aberrations and distortions [10, 11] is also possible. To date, however, substantial impact on industrial processes and applications is still in the future.

7.3 Thermal Transport Studied with Femtosecond Light Pulses

Determining material properties like heat transport and elasticity is of fundamental interest in industry, especially in understanding the unexpected

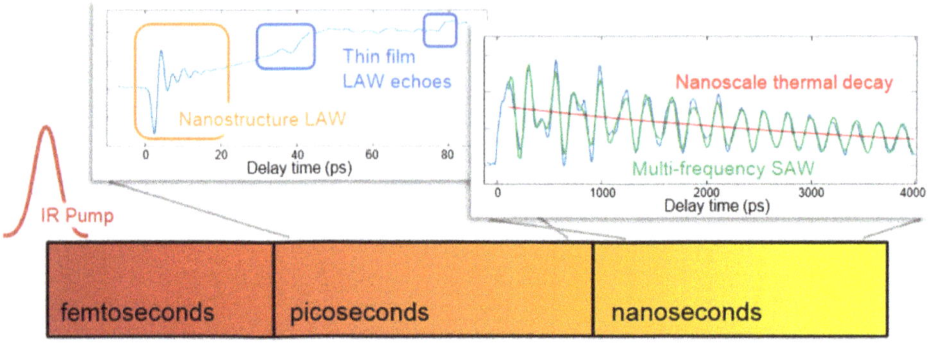

Fig. 2. Dynamics observed in pump-probe diffraction of nanoarrays after femtosecond excitation. Picosecond timescales are marked by LAWs in the nanostructure and, in the case shown, a thin film in the nanostructure stack. For nanosecond timescales, multifrequency SAWs waves are seen, riding on a longer timescale thermal decay of the overall nanostructure [23].

ways in which these materials behave as feature sizes are scaled down on semiconductor chips. Using similarly generated EUV light from a femtosecond laser, as used in CDI discussed above, both the thermal transport in nanoscale structures and material elasticity in thin films can be probed in a "pump-probe" geometry (Fig. 2). This pump being incident on a nanostructure array serves to (1) heat the nanostructures, expanding them from their original size and (2) launch oscillations in the nanostructures, or surface and longitudinal acoustic waves (SAWs and LAWs, respectively), similar to the vibrations of a drum head upon being struck. The femtosecond EUV pulse then "probes" by diffraction from the pumped nanostructure and the diffracted pattern is collected. The shift in diffracted spectra in the pumped case versus unpumped allows one to extract the short-term (LAW oscillations) and long-term (thermal decay and SAW) dynamics. These dynamics range from picosecond to nanosecond timescales.

In this case, the ultrafast driving laser serves to generate the probing light, and the EUV gives the advantage of a larger, more easily observed offset in the diffraction. Heat dissipation from features down to 20 nm have been well characterized, observing the transition between multiple heat transport regimes [12]. Additionally, thin films down to 50 nm thickness, possessing varying elasticities, have been measured [13, 14].

7.4 X-ray Free-Electron Lasers (XFEL)

Another emerging source of ultrashort pulses of light are X-ray free-electron lasers or XFELs. An example is the Linac Coherent Light Source (LCLS) at the Stanford Linear Accelerator Laboratory [12, 13]. A high-energy pulsed beam of electrons passes through an undulator, a periodic linear array of dipole magnets that forces the electrons to execute an oscillatory trajectory in a plane perpendicular to their injected direction. The accelerated electrons radiate coherently in the X-ray. The LCLS FEL is tunable from 1.5 to 15 Å based upon the energy of the electron beam which can vary from 4.3 to 14 GeV. The pulsewidth is as short as 40 fs but can be made shorter with lower photon flux. A significant aspect of this type of ultrashort pulsed laser is that it is now possible to capture an entire X-ray diffraction pattern [14] with a single laser pulse. While such a pulse would possess an extremely high photon flux that would, in essence, destroy the sample, the ultrashort nature of the X-ray pulse captures the diffraction pattern before the atoms within the sample can move. This is a powerful approach to capturing diffraction patterns from delicate biological and radiation-sensitive nanostructures.

XFELs have also been considered as sources for EUV lithography [15, 16]. Expectations of as much as 5 kW of 13.5 nm power would be available as input into an EUV stepper. Present EUV stepper technology utilizes large CO_2 lasers to transform metal droplets into plasmas that radiate over a wide range of wavelengths over all space. The energetic plasma that forms must be controlled, but eventually damages reflective optical components in the stepper. Alternatively, an XFEL source is clean and the light is highly directional, and would be an excellent candidate for lithography [17, 18], assuming such a laser could provide sufficient flux at 13.5 nm in a cost-effective manner.

Figure 3 shows the undulator hall where electron bunches travel through alternating magnetic regions that drive their oscillatory motion; the bunch spontaneously emits radiation that is enhanced at the wavelength dictated by the geometry of the undulator itself.

Figure 4 shows the tremendous peak brilliance and power available from FELs in comparison with other light sources such as storage rings, table-top lasers, and high-harmonic generation.

7.5 Synchrotron Facilities

Synchrotron facilities such as that shown in Fig. 5 of the recently completed National Synchrotron Light Source (NSLS) II [19] are not typically considered sources of ultrashort pulses but because of their extraordinarily wide range of tunability combined with relatively short pulses of light, these sources are of considerable interest and utility. Some of the

Fig. 3. LCLS FEL undulator hall.

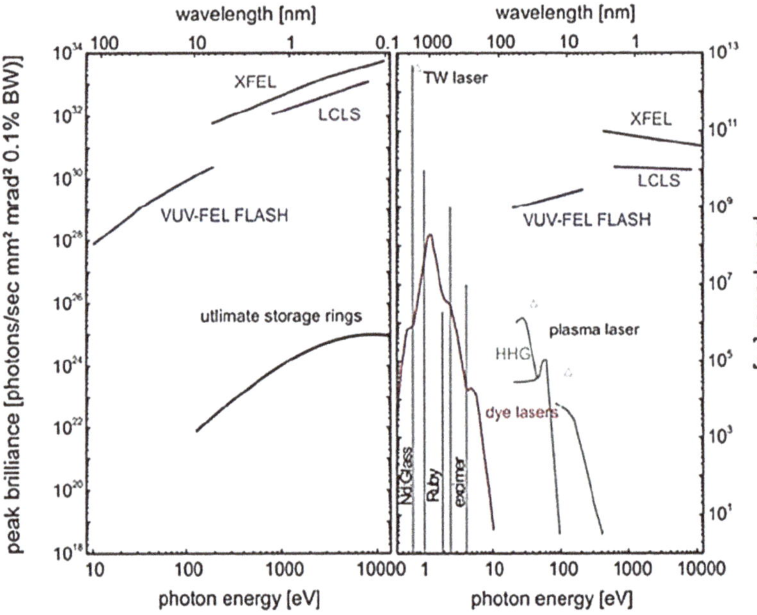

Fig. 4. Peak brilliance and power of FEL facilities compared with storage rings and table-top laser-based sources (reference: https://science.energy.gov/~/meida/bes/pdf/reports/files/ngps_rpt_print.pdf (2008)).

Fig. 5. Bird's-eye view of the NSLS II at the Brookhaven National Lab.

key parameters of the NSLS II are the wide range of tunability from the IR (down to 100 μeV) to very hard X-rays of 300 keV (see, for example (https://www.bnl.gov/ps/docs/pdf/SourceProperties.pdf). The root mean square (RMS) pulse length can be as short as 15–30 ps. Synchrotrons have been used for a wide array of industrial applications including early pioneering work on X-ray and EUV lithography and X-ray spectroscopies such as XPS, X-ray absorption-fine structure experiments, and many more. NSLS II is a third-generation source with extraordinary brightness as shown in Fig. 6. While XFELs tend to be single user with secondary experiments utilizing transmitted light from the first, synchrotrons are fundamentally multi-user sources with multiple beamlines and experiments arrayed around the ring, typically at the output of turning or wiggler magnet sections.

7.6 Final Comments

This book has been written with the intention of providing a glimpse into the types of industrial applications that can be addressed with ultrashort pulses of light. Ultrafast optical pulses, typically in the femtosecond regime, but more recently reaching into the attosecond time scale with high harmonics, are finding ever greater utility. Combined with nonlinear optical processes,

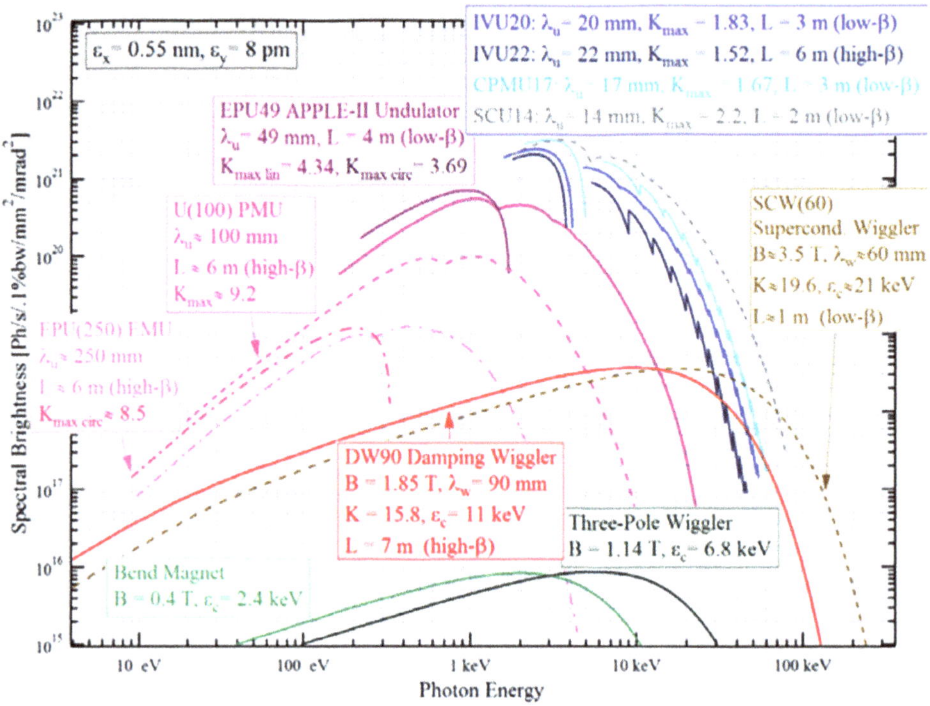

Fig. 6. Brightness as a function of photon energy for various NSLS-II radiation sources, at 3 GeV and 500 mA (https://www.bnl.gov/ps/docs/pdf/SourceProperties.pdf-2010).

ultrashort laser pulses are employed additionally in photonics, chemistry, medical, and biological applications.

The focus in this work is on the utilization of intense ultrashort pulses of laser light for spectroscopic materials analysis and materials modification through ablation and machining. The list of applications and studies is ever growing, and it is the authors' general sense that this expansion will continue for many years to come as the actual laser sources themselves become more compact, lower cost and much more reliable. As can be seen in this very brief survey chapter and from the preceding six chapters describing applications, the interaction of light with matter can occur on the small scale with individual lasers on an optical table in a lab to the very grand scale in XFELs and enormous synchrotron facilities at national laboratories. Our knowledge of light–matter interactions and new surprises that present themselves almost daily in both materials innovation and dynamics at ever expanding time and energy scales will continue to provide avenues for scientific investigation for many decades to come.

References

[1] V. Bakshi, *EUV Lithography*, Vol. 178. SPIE Press Bellingham, 2009.
[2] V. Banine and R. Moors, "Plasma sources for EUV lithography exposure tools," *J. Phys. D. Appl. Phys.*, vol. 37, no. 23, p. 3207, 2004.
[3] T. Gershon, Y. S. Lee, P. Antunez, R. Mankad, S. Singh, D. Bishop, O. Gunawan, M. Hopstaken, and R. Haight, "Photovoltaic materials and devices based on the alloyed kesterite absorber (AgxCu1–x) 2ZnSnSe4," *Adv. Energy Mater.*, 2016.
[4] B. W. J. McNeil and N. R. Thompson, "X-ray free-electron lasers," *Nat Phot.*, vol. 4, no. 12, pp. 814–821, 2010.
[5] A. Rundquist, C. G. Durfee, Z. Chang, C. Herne, S. Backus, M. M. Murnane, and H. C. Kapteyn, "Phase-matched generation of coherent soft X-rays," *Science*, vol. 280, no. 5368, pp. 1412–1415, 1998.
[6] R. A. Bartels, A. Paul, H. Green, H. C. Kapteyn, M. M. Murnane, S. Backus, I. P. Christov, Y. Liu, D. Attwood, and C. Jacobsen, "Generation of spatially coherent light at extreme ultraviolet wavelengths," *Science*, vol. 297, no. 5580, pp. 376–378, 2002.
[7] X. Zhang, A. R. Libertun, A. Paul, E. Gagnon, S. Backus, I. P. Christov, M. M. Murnane, H. C. Kapteyn, R. A. Bartels, and Y. Liu, "Highly coherent light at 13 nm generated by use of quasi-phase-matched high-harmonic generation," *Opt. Lett.*, vol. 29, no. 12, pp. 1357–1359, 2004.
[8] J. Miao, T. Ishikawa, I. K. Robinson, and M. M. Murnane, "Beyond crystallography: Diffractive imaging using coherent X-ray light sources," *Science*, vol. 348, no. 6234, p. 530 LP–535, 2015.
[9] M. D. Seaberg, D. E. Adams, E. L. Townsend, D. A. Raymondson, W. F. Schlotter, Y. Liu, C. S. Menoni, L. Rong, C. Chen, J. Miao, H. C. Kapteyn, and M. M. Murnane, "Ultrahigh 22 nm resolution coherent diffractive imaging using a desktop 13 nm high harmonic source," *Opt. Exp.* vol. 19, no. 23, pp. 7235–7239, 2011.
[10] D. G. Lee, J. J. Park, J. H. Sung, and C. H. Nam, "Wave-front phase measurements of high-order harmonic beams by use of point-diffraction interferometry," *Opt. Lett.*, vol. 28, no. 6, pp. 480–482, 2003.
[11] E. J. Takahashi, Y. Nabekawa, and K. Midorikawa, "Low-divergence coherent soft x-ray source at 13 nm by high-order harmonics," *Appl. Phys. Lett.*, vol. 84, no. 1, pp. 4–6, 2004.
[12] K. Hoogeboom-Pot, N. Hernandez-Charpak, T. Frazer, X. Gu, E. Turgut, E. Anderson, W. Chao, J. Shaw, R. Yang, M. M. Murnane, H. C. Kapteyn, and D. Nardi, "Mechanical and ther- mal properties of nanomaterials at sub-50 nm dimensions characterized using coherent EUV beams," *Proc SPIE Adv. Lithogr.*, vol. 9424, 2015.
[13] Q. Li, K. Hoogeboom-Pot, D. Nardi, C. Deeb, S. Kind, M. Tripp, E. Anderson, M. M. Murnane, and H. C. Kapteyn, "Characterization of ultrathin films by laser-induced sub-picosecond photoacoustics with coherent extreme ultraviolet detection," *Proc. SPIE*, vol. 8324, p. 83241P, 2012.
[14] D. Nardi, K. Hoogeboom-Pot, N. Hernandez-Charpak, M. Tripp, S. King, E. Anderson, M. M. Murnane, and H. C. Kapteyn, "Probing limits of acoustic nanometrology using coherent extreme ultraviolet light," *Proc SPIE Adv. Lithogr.*, p. 86810N, 2013.
[15] P. Emma, "First lasing of the LCLS X-ray FEL at 1.5 Å," *Proc. PAC09, Vancouver, to be Publ. http//accelconf.web.Cern.ch/AccelConf*, 2009.
[16] P. Emma, R. Akre, J. Arthur, R. Bionta, C. Bostedt, J. Bozek, A. Brachmann, P. Bucksbaum, R. Coffee, and F.-J. Decker, "First lasing and operation of an

ångstrom-wavelength free-electron laser," *Nat. Photonics*, vol. 4, no. 9, pp. 641–647, 2010.

[17] H. N. Chapman, P. Fromme, A. Barty, T. A. White, R. A. Kirian, A. Aquila, M. S. Hunter, J. Schulz, D. P. DePonte, and U. Weierstall, "Femtosecond X-ray protein nanocrystallography," *Nature*, vol. 470, no. 7332, pp. 73–77, 2011.

[18] W. Ackermann, *et al.*, "Operation of a free-electron laser from the extreme ultraviolet to the water window," *Nat, Phot.*, vol. 1, no. 6, pp. 336–342, 2007.

[19] Y. Socol, G. N. Kulipanov, A. N. Matveenko, O. A. Shevchenko, and N. A. Vinokurov, "Compact 13.5-nm free-electron laser for extreme ultraviolet lithography," *Phys. Rev. Spec. Top. Beams*, vol. 14, no. 4, p. 40702, 2011.

[20] T. Shintake, H. Tanaka, T. Hara, T. Tanaka, K. Togawa, M. Yabashi, Y. Otake, Y. Asano, T. Bizen, and T. Fukui, "A compact free-electron laser for generating coherent radiation in the extreme ultraviolet region," *Nat. Photonics*, vol. 2, no. 9, pp. 555–559, 2008.

[21] E. R. Hosler, O. R. Wood, W. A. Barletta, P. J. S. Mangat, and M. E. Preil, "Considerations for a free-electron laser-based extreme-ultraviolet lithography program," in *SPIE Advanced Lithography*, p. 94220D–94220D, 2015.

[22] F. Willeke, "Commissioning of NSLS-II," in *Proc of IPAC15, International Particle Accelerator Conference, Virginia, USA (JACoW, 2015)*, p. 11, 2015.

[23] K. Hoogeboom-Pot, "Uncovering new thermal and mechanical behavior at the nanoscale using coherent extreme ultraviolet light", PhD Thesis, University of Colorado Boulder, 2015.

Index

World Scientific Series in Materials and Energy

(Continuation of series card page)

Vol. 8 *Thin Films on Silicon: Electronic and Photonic Applications*
edited by Vijay Narayanan (IBM Thomas J Watson Research Center, USA), Martin M. Frank (IBM Thomas J Watson Research Center, USA) and Alexander A. Demkov (The University of Texas at Austin, USA)

Vol. 7 *The WSPC Reference on Organic Electronics: Organic Semiconductors (In 2 Volumes)*
edited by Jean-Luc Brédas (King Abdullah University of Science & Technology, Saudi Arabia & Georgia Institute of Technology, USA) and Seth R. Marder (Georgia Institute of Technology, USA)

Vol. 6 *Handbook of Solid State Batteries (Second Edition)*
edited by Nancy J. Dudney (Oak Ridge National Laboratory, USA), William C. West (Nagoya University, Japan) and Jagjit Nanda (Oak Ridge National Laboratory, USA)

Vol. 5 *Handbook of Green Materials: Processing Technologies, Properties and Applications (In 4 Volumes)*
edited by Kristiina Oksman (Luleå University of Technology, Sweden), Aji P. Mathew (Luleå University of Technology, Sweden), Alexander Bismarck (Vienna University of Technology, Austria), Orlando Rojas (North Carolina State University, USA), and Mohini Sain (University of Toronto, Canada)

Vol. 4 *Materials for Biofuels*
edited by Arthur J. Ragauskas (Georgia Institute of Technology, USA)

Vol. 3 *The World Scientific Handbook of Energy*
edited by Gerard M. Crawley (University of South Carolina, USA)

Vols. 1 & 2 *Handbook of Instrumentation and Techniques for Semiconductor Nanostructure Characterization*
edited by Richard Haight (IBM TJ Watson Research Center, USA), Frances M. Ross (IBM TJ Watson Research Center, USA), James B. Hannon (IBM TJ Watson Research Center, USA) and Leonard C. Feldman (Rutgers University, USA)

www.ingramcontent.com/pod-product-compliance
Ingram Content Group UK Ltd.
Pitfield, Milton Keynes, MK11 3LW, UK
UKHW052114170926
8624IPUK00026B/386

9 789814 569002